METHODS IN MOLECULAR BIOLOGY

T0335408

Series Editor
John M. Walker
School of Life and Medical Sciences
University of Hertfordshire
Hatfield, Hertfordshire, AL10 9AB, UK

Biotechnology of Plant Secondary Metabolism

Methods and Protocols

Edited by

Arthur Germano Fett-Neto

Federal University of Rio Grande do Sul, Porto Alegre, RS, Brazil

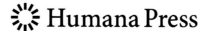

Editor
Arthur Germano Fett-Neto
Federal University of Rio Grande do Sul
Porto Alegre, RS, Brazil

ISSN 1064-3745   ISSN 1940-6029   (electronic)
Methods in Molecular Biology
ISBN 978-1-4939-3391-4   ISBN 978-1-4939-3393-8   (eBook)
DOI 10.1007/978-1-4939-3393-8

Library of Congress Control Number: 2015957398

Springer New York Heidelberg Dordrecht London

Cover Illustration: Seedling of Quillaja brasiliensis (A. St.-Hill. & Tul.) Mart., a tree whose leaves yield immunoadjuvant saponins.

Printed on acid-free paper

Humana Press is a brand of Springer
Springer Science+Business Media LLC New York is part of Springer Science+Business Media (www.springer.com)

Preface

Humans depend on plants not only for food and oxygen but also for medicines, energy, fiber, and building materials. Plant secondary metabolites, in particular, are at center stage as a major source of pharmaceuticals, products for the chemical and food industry, agrochemicals, and special biofuels. Proper manipulation of plant physiology and molecular biology, by means of an array of biotechnological strategies, can promote large gains in yield of desired products. In this book, up-to-date representative methods for improved production of secondary metabolites of economic interest are described in detail, along with tips to avoid common pitfalls in these approaches. An effort was made to include examples applied to the production of different secondary metabolite classes, using field and laboratory methods, whole-plant and cell/organ culture systems, as well as environmental and genetic transformation-based modulation of biochemical pathways. In agreement with current trends in plant secondary metabolism research, a focus was given on cell and tissue specific metabolism, metabolite transport, microRNA-based technology, heterologous systems expression of enzymes and pathways leading to products of interest, as well as applications using both model and non-model plant species. It is our expectation that this volume may aid scientists of interdisciplinary fields (plant science, plant physiology, pharmacy, molecular biology, biochemistry, bioengineering, and forestry) in reaching the major goal of producing fine plant biochemicals in sustainable and efficient fashion, in order to minimize impacts to the environment and provide the required quantities of these commodities to industry.

Porto Alegre, Brazil   *Arthur G. Fett-Neto*

Contents

Contributors

MÁRCIA R. DE ALMEIDA • *Center for Biotechnology, Federal University of Rio Grande do Sul, Porto Alegre, RS, Brazil*

CARLA J.S. BARBER • *National Research Council of Canada, Saskatoon, SK, Canada*

SARA BETTENCOURT • *Instituto de Investigação e Inovação em Saúde, Universidade do Porto, Porto, Portugal; IBMC – Instituto de Biologia Molecular e Celular, Universidade do Porto, Porto, Portugal*

BASDEO BHAGWAT • *Pacific Agri-Food Research Center, Agriculture and Agri-Food Canada, Summerland, BC, Canada*

DULAL BORTHAKUR • *Department of Molecular Biosciences and Bioengineering, University of Hawaii at Manoa, Honolulu, HI, USA*

INÊS CARQUEIJEIRO • *Departamento de Biologia, Faculdade de Ciências, Universidade do Porto, Porto, Portugal; IBMC – Instituto de Biologia Molecular e Celular, Universidade do Porto, Porto, Portugal; Instituto de Investigação e Inovação em Saúde, Universidade do Porto, Porto, Portugal*

MING CHI • *Pacific Agri-Food Research Center, Agriculture and Agri-Food Canada, Summerland, BC, Canada; College of Forestry, Northwest A & F University, Yangling, Shaanxi, China*

FERNANDA DE COSTA • *Plant Physiology Laboratory, Department of Botany, Federal University of Rio Grande do Sul (UFRGS), Porto Alegre, RS, Brazil; Center for Biotechnology, Federal University of Rio Grande do Sul (UFRGS), Porto Alegre, RS, Brazil*

PATRICK S. COVELLO • *National Research Council of Canada, Saskatoon, SK, Canada*

PATRÍCIA DUARTE • *Instituto de Investigação e Inovação em Saúde, Universidade do Porto, Porto, Portugal; IBMC – Instituto de Biologia Molecular e Celular, Universidade do Porto, Porto, Portugal*

ANITA EMMERSTORFER-AUGUSTIN • *ACIB – Austrian Centre of Industrial Biotechnology, Graz, Austria*

ARTHUR GERMANO FETT-NETO • *Plant Physiology Laboratory, Center for Biotechnology, Federal University of Rio Grande do Sul (UFRGS), Porto Alegre, RS, Brazil; Department of Botany, UFRGS, Porto Alegre, RS, Brazil*

THANISE NOGUEIRA FÜLLER • *Plant Physiology Laboratory, Center for Biotechnology, Federal University of Rio Grande do Sul (UFRGS), Porto Alegre, RS, Brazil; Department of Botany, UFRGS, Porto Alegre, RS, Brazil*

ANDY GANAPATHI • *Department of Biotechnology and Genetic Engineering, School of Biotechnology, Bharathidasan University, Tiruchirappalli, Tamil Nadu, India*

HERNÂNI GERÓS • *Centro de Investigação e de Tecnologias Agro-Ambientais e Biológicas (CITAB-UM), Braga, Portugal; Grupo de Investigação em Biologia Vegetal Aplicada e Inovação Agroalimentar, Departamento de Biologia, Universidade do Minho, Braga, Portugal*

BARBARA ANN HALKIER • *DynaMo center of Excellence, Department of Plant and Environmental Sciences, Faculty of Science, University of Copenhagen, Frederiksberg C, Denmark*

DIANWEI HAN • *Department of Computer Science, University of Kentucky, Lexington, KY, USA*

MORTEN EGEVANG JØRGENSEN • *DynaMo Center, Department of Plant and Environmental Sciences, Faculty of Science, University of Copenhagen, Frederiksberg C, Denmark*

EDUARDO KIYOTA • *Departamento de Biologia Vegetal-IB, Universidade Estadual de Campinas, Campinas, SP, Brazil*

JÚLIO CÉSAR DE LIMA • *Molecular Genetics Laboratory, University of Passo Fundo, Passo Fundo, RS, Brazil*

SVEND ROESEN MADSEN • *DynaMo Center of Excellence, Department of Plant and Environmental Sciences, Faculty of Science, University of Copenhagen, Frederiksberg C, Denmark*

MARKANDAN MANICKAVASAGAM • *Department of Biotechnology and Genetic Engineering, School of Biotechnology, Bharathidasan University, Tiruchirappalli, Tamil Nadu, India*

VIVIANA MARTINS • *Instituto de Investigação e Inovação em Saúde, Universidade do Porto, Porto, Portugal; IBMC – Instituto de Biologia Molecular e Celular, Universidade do Porto, Porto, Portugal; Centro de Investigação e de Tecnologias Agro-Ambientais e Biológicas (CITAB-UM), Braga, Portugal; Grupo de Investigação em Biologia Vegetal Aplicada e Inovação Agroalimentar, Departamento de Biologia, Universidade do Minho, Braga, Portugal*

PAULO MAZZAFERA • *Departamento de Biologia Vegetal-IB, Universidade Estadual de Campinas, Campinas, SP, Brazil*

VISHAL SINGH NEGI • *Department of Molecular Biosciences and Bioengineering, University of Hawaii at Manoa, Honolulu, HI, USA*

HENRIQUE NORONHA • *Centro de Investigação e de Tecnologias Agro-Ambientais e Biológicas (CITAB-UM), Braga, Portugal; Grupo de Investigação em Biologia Vegetal Aplicada e Inovação Agroalimentar, Departamento de Biologia, Universidade do Minho, Braga, Portugal*

HUSSAM HASSAN NOUR-ELDIN • *DynaMo Center of Excellence, Department of Plant and Environmental Sciences, Faculty of Science, University of Copenhagen, Frederiksberg C, Denmark*

HARALD PICHLER • *ACIB – Austrian Centre of Industrial Biotechnology, Graz, Austria; Institute of Molecular Biotechnology, NAWI Graz, Graz University of Technology, Graz, Austria*

PARESHKUMAR T. PUJARA • *National Research Council of Canada, Saskatoon, SK, Canada*

DARWIN W. REED • *National Research Council of Canada, Saskatoon, SK, Canada*

DIANA RIBEIRO • *Departamento de Biologia, Faculdade de Ciências, Universidade do Porto, Porto, Portugal; Instituto de Investigação e Inovação em Saúde, Universidade do Porto, Porto, Portugal; IBMC – Instituto de Biologia Molecular e Celular, Universidade do Porto, Porto, Portugal; Departamento de Biologia, Universidade do Minho, Braga, Portugal*

KELLY C.S. RODRIGUES-CORRÊA • *Regional Integrated University of Alto Uruguai and Missões (URI-FW), Frederico Westphalen, RS, Brazil*

FLÁVIA CAMILA SCHIMPL • *Departamento de Biologia Vegetal-IB, Universidade Estadual de Campinas, Campinas, SP, Brazil*

NATESAN SELVARAJ • *Department of Botany, Periyar E. V. R College (Autonomous), Tiruchirappalli, Tamil Nadu, India*

GANESHAN SIVANANDHAN • *Molecular Genetics and Genomics Laboratory, Department of Horticulture, College of Agriculture and Life Sciences, Chungnam National University, Daejeon, South Korea*

MARIANA SOTTOMAYOR • *Departamento de Biologia, Faculdade de Ciências, Universidade do Porto, Porto, Portugal; Instituto de Investigação e Inovação em Saúde, Universidade do Porto, Porto, Portugal; IBMC – Instituto de Biologia Molecular e Celular, Universidade do Porto, Porto, Portugal*

MARTINA V. STRÖMVIK • *Department of Plant Science, McGill University, Sainte-Anne-de-Bellevue, QC, Canada*

HAIFENG TANG • *Provincial State Key Laboratory of Wheat and Maize Crop Science, Henan Agricultural University, Zhengzhou, China*

GUILIANG TANG • *Provincial State Key Laboratory of Wheat and Maize Crop Science, Henan Agricultural University, Zhengzhou, China; Department of Biological Sciences, Michigan Technological University, Houghton, MI, USA*

YU XIANG • *Pacific Agri-Food Research Center, Agriculture and Agri-Food Canada, Summerland, BC, Canada*

Chapter 1

Elicitation Approaches for Withanolide Production in Hairy Root Culture of *Withania somnifera* (L.) Dunal

Ganeshan Sivanandhan, Natesan Selvaraj, Andy Ganapathi, and Markandan Manickavasagam

Abstract

Withania somnifera (L.) Dunal is a versatile medicinal plant extensively utilized for production of phytochemical drug preparations. The roots and whole plants are traditionally used in Ayurveda, Unani, and Siddha medicines, as well as in homeopathy. Several studies provide evidence for an array of pharmaceutical properties due to the presence of steroidal lactones named "withanolides." A number of research groups have focused their attention on the effects of biotic and abiotic elicitors on withanolide production using cultures of adventitious roots, cell suspensions, shoot suspensions, and hairy roots in large-scale bioreactor for producing withanolides. This chapter explains the detailed procedures for induction and establishment of hairy roots from leaf explants of *W. somnifera*, proliferation and multiplication of hairy root cultures, estimation of withanolide productivity upon elicitation with salicylic acid and methyl jasmonate, and quantification of major withanolides by HPLC. The protocol herein described could be implemented for large-scale cultivation of hairy root biomass to improve withanolide production.

Key words *Agrobacterium rhizogenes*, Hairy root culture, Salicylic acid, Methyl jasmonate exposure time, Inoculum mass, Time course study, Withanolides, HPLC

1 Introduction

Withania somnifera (L.) Dunal (*Solanaceae*), commonly known as "Ashwagandha"/"Indian ginseng," is a highly valued medicinal plant in Indian Ayurvedic and African traditional medicine [1]. The medicinal importance of *Withania somnifera* (*Solanaceae*) is attributed to the presence of steroidal lactones named "withanolides" [2, 3]. The pharmaceutically important phytochemicals found in this plant species are withanolide A, withaferin A, and withanone (major constituents; Fig. 1) [4]. Recently, the biosynthesis of withanolides has been updated with some information in *W. somnifera* (Fig. 2) [3]. Medicinal importance of withanolides comprises physiological and metabolic restoration, antiarthritic, antiageing, nerve stimulant, cognitive function improvement in

Arthur Germano Fett-Neto (ed.), *Biotechnology of Plant Secondary Metabolism: Methods and Protocols*, Methods in Molecular Biology, vol. 1405, DOI 10.1007/978-1-4939-3393-8_1, © Springer Science+Business Media New York 2016

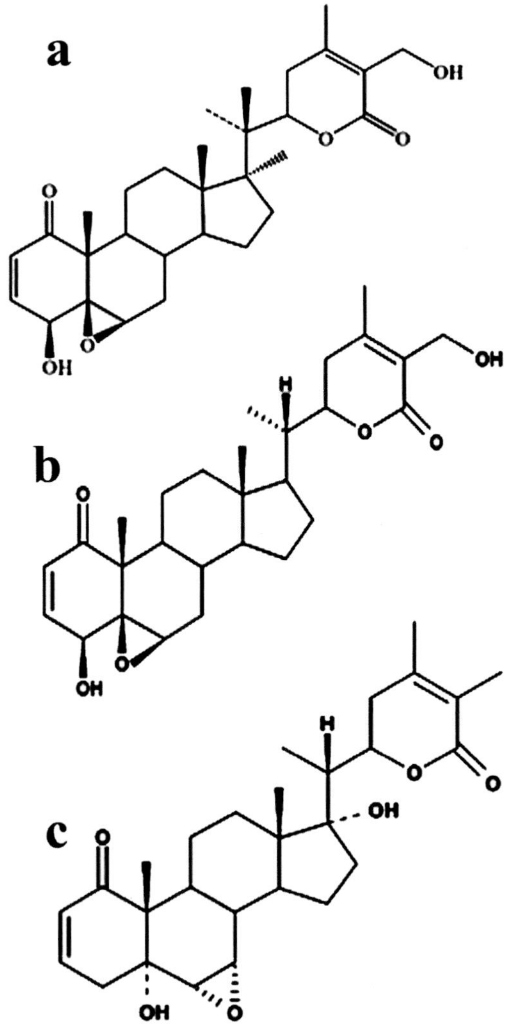

Fig. 1 Structure of withanolides biosynthesized in *W. somnifera*. (**a**) Withanolide A, (**b**) withaferin A, (**c**) withanone

the elderly, and recovery from neurodegenerative disorders [5]. The traditional cultivation methods of *W. somnifera* for withanolide drug preparation have been limited by a range of issues such as biotic and abiotic environmental factors, unpredictability of bioactive components, and lack of purity and standardized plant raw material for phytochemical analysis [6]. Moreover, these methods are time consuming, laborious, and not able to sustain the current ashwagandha global market requirement [7]. The requirement of dried plant material for withanolide production in India has been estimated at about 9127 t, whereas the annual production is of about 5905 t [8]. At the international level, there has been an ever-increasing demand for *W. somnifera* in larger quantities [7].

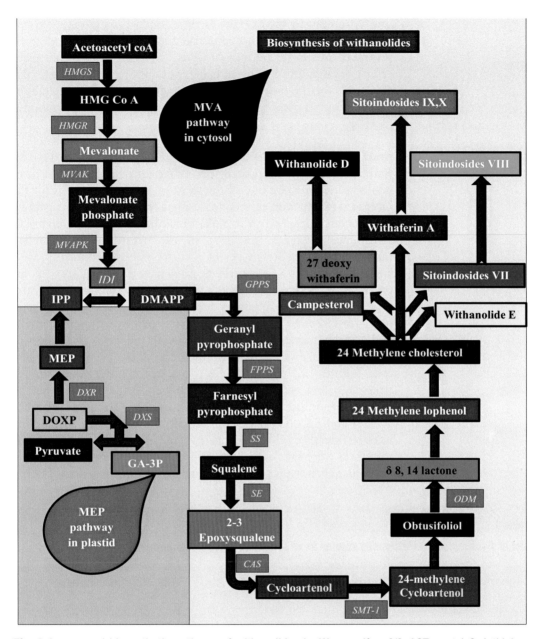

Fig. 2 A proposed biosynthetic pathway of withanolides in *W. somnifera* [4]. ACT-*acetyl-CoA thiolase*, HMGS-3-hydroxy3-*methylglutarylCo-A synthase*, HMGR-*3-hydroxy3-methylglutaryl CoA reductase*, MVAK-*mevalonate kinase*, MVAPK-mevalonate*phosphate kinase*, MVAPPD-*mevalonate pyrophosphate decarboxylase*, GPPS-*geranyl pyrophosphate synthase*, FPP-*farnesyl pyrophosphate synthase*, SS-*squalene synthase*, SE-*squalene epoxidase*, CAS-*cycloartenol synthase*, BAS-β-amyrin*synthase*, SMT-1-*sterol methyl transferase* 1, ODM-*obtusifoliol demethylase* (*long arrows* represent multiple steps)

As traditional cultivation of Indian ginseng is time consuming and laborious and the demand of the dried product is increasing, cell and organ cultures have been used for production of secondary metabolites; however, yields of withanolide metabolites have been low. Attempts to increase levels of secondary metabolites in cell and organ cultures should be pursued as these metabolites can be produced in shorter periods of time [7, 9]. Among various plant cell or organ culture systems, hairy root culture is one of the valuable tools for the biosynthesis of secondary metabolites, metabolic engineering studies, and production of root-derived compounds [1, 10].

Earlier reports documented relatively low levels of withanolides from hairy root cultures of *W. somnifera* (Table 1) [1, 11–16]. Radman et al. [17] reported that exogenous addition of elicitors was considered to be one of the most promising strategies for the increased production of secondary metabolites. The elicitor molecule in culture interacts with a plant membrane receptor which activates specific genes, resulting in the synthesis of secondary metabolites. Of various types of elicitors, methyl jasmonate (MeJ) and salicylic acid (SA) have been proven as active elicitors for the production of secondary metabolites in plant cell/organ cultures [7, 18]. MeJ is a ubiquitous signaling molecule which mediates plant responses to environmental stress such as wounding, insect, and pathogen attack [19] often enhancing the yields of a variety of secondary metabolites [20–22], including those produced in hairy root cultures [23]. Similarly, SA is one of the most widely studied stress-signaling molecules and it influences plant resistance to pathogens and other stress factors [7, 24, 25]. Though the impact of MeJ and SA on enhanced production of secondary metabolites has been clearly established in hairy root cultures of several taxa, the responses were species specific, and there is no universal effect of a particular elicitor on different plants or cell culture systems [17].

Table 1

List of *Agrobacterium rhizogenes* studies in *W. somnifera* by different authors

Authors	Explant	Withanolides reported
Pawar and Maheswari [11]	Stem, hypocotyl, and leaves	–
Kumar et al. [12]	Cotyledon and leaves	–
Bandyopadhyay et al. [13]	Leaves	Withaferin A and withanolide D
Murthy et al. [14]	Cotyledon, leaves, stem, root, hypocotyl, and cotyledonary node	Withanolide A
Saravanakumar et al. [15]	Leaf, petiole, and internode	Withaferin A
Sivanandhan et al. [1]	Leaves	Withanolide A, withanone, and withaferin A
Sivanandhan et al. [16]	Leaves	Withanolide A, withanone, and withaferin A

We have investigated a variety of biotic and abiotic elicitors for the enhancement of major and minor withanolide production in multiple shoots, adventitious roots and hairy root cultures of *W. somnifera*, and a number of effective biotic and abiotic elicitors were identified [1, 2, 4, 6, 7, 9, 16]. Among the different organ cultures as well as elicitors tested, we showed that SA elicitation can be used for withanolide production (42–58-fold) in hairy root culture of *W. somnifera*. In the present protocol, the use of elicitors for withanolide production in hairy root culture using 250 mL and 3 L shake-flask culture systems in *W. somnifera* is described.

2 Materials

2.1 Explant Preparation

1. Mature seeds of *W. somnifera* (*see* **Note 1**).
2. 2.5 % (v/v) commercial bleach Teepol® (5.25 % sodium hypochlorite).
3. 0.1 % (w/v) $HgCl_2$.
4. Sterilized distilled water.
5. Sterilized cotton moistened with sterile water.

2.2 Hairy Root Induction and Proliferation

1. *Agrobacterium rhizogenes* R1000 (*see* **Note 2**).
2. LB medium (Himedia, Mumbai, India).
3. Rotary shaker (at 180, 120, and 80 rpm).
4. Murashige and Skoog [MS] [26] medium stock solutions: Stocks may be prepared using individual chemicals (Table 2). Store at 4 °C (*see* **Note 3**).
5. Acetosyringone [150 μM] (Sigma, St. Louis, USA).
6. DMSO.
7. MES buffer [20 mM] (pH 5.4).
8. Hypodermic needle.
9. Sterile filter paper.
10. Sucrose.
11. 0.2 % (w/v) Phytagel (Sigma, St. Louis, USA) or 0.8 % (w/v) agar.
12. Cefotaxime (Alkime Laboratory, Mumbai, India).

2.3 Assessment of Elicitation Conditions on Biomass Accumulation and Withanolide Production

1. MeJ and SA (methyl jasmonate and salicylic acid) (Sigma, St. Louis, USA).
2. 99.9 % Ethanol and Milli-Q water.
3. Petri dishes (15 mm × 100 mm).
4. 0.20 μm PTFE membrane filters (Pall India Pvt. Ltd., Mumbai, India).

Table 2

MS medium acomposition

Chemical constituents	Concentration (mg/L)
Macro	
$(NH_4)NO_3$	1650
KNO_3	1900
$CaCl_2 \cdot 2H_2O$	440
$MgSO_4 \cdot 7H_2O$	370
KH_2PO_4	170
Micro	
$MnSO_4 \cdot 4H_2O$	22.3
$ZnSO_4 \cdot 4H_2O$	8.6
H_3BO_3	6.2
Minor	
$Na_2Mo_4 \cdot 2H_2O$	0.25
$CuSO_4 \cdot 5H_2O$	0.025
$CoCl_2 \cdot 6H_2O$	0.025
Iron[a]	
Na_2 EDTA	37.3
$FeSO_4 \cdot 7H_2O$	27.8
KI	0.83
Myoinositol	100
Vitamins	
Thiamine HCl	0.1
Pyridoxine HCl	0.5
Nicotinic acid	0.5
Glycine	2.0
Sucrose	As per requirement

After dissolving all the stocks in enough distilled water, make it up to 1 L, adjust the pH to 5.6–5.8 (add 2 g/L Phytagel or 0.8 % agar for solid medium), and autoclave for 20 min at 120 °C

[a]Add Na_2 EDTA in water at moderate temperature (approx. 30 °C) for 30 min, up to the point that this solution becomes transparent. Then, slowly add $FeSO_4 \cdot 7H_2O$ for iron stock preparation

2.4 Withanolide
Extraction and HPLC
Analysis

1. Hairy root biomass.
2. Mortar and pestle.
3. Liquid nitrogen.
4. HPLC-grade methanol (Merck Chemicals, India) (*see* **Note 4**).
5. HPLC-grade water (Merck Chemicals, India) (*see* **Note 4**).
6. Isocratic mobile phase of methanol and water (65:35 v/v).
7. Sonicator (Sae Han Ultrasonic Co, Korea) (*see* **Note 5**).
8. Oven (Biojenik Pvt. Ltd., Chennai, India) (*see* **Note 6**).
9. Vacuum desiccator (Tarsons Products Pvt. Ltd., Kolkata, India).
10. Membrane syringe filters (0.20 μm pore) for HPLC samples.
11. HPLC ([Waters HPLC, Vienna, Austria] equipped with a PDA detector and a reverse-phase Luna® column C18 [5 μm; 30×4.6 mm]).
12. Withanolide A standard (Chromadex Inc., Laguna Hills, CA, USA). Withaferin A, and withanone standards (Natural Remedies, Bengaluru, Karnataka, India) (*see* **Note 7**).

3   Methods

3.1 Explant
Preparation

1. Sterilize the mature seeds by immersion in 2.5 % (v/v) bleach and in 0.1 % $HgCl_2$ for 2 min (*see* **Note 8**) (Fig. 3).
2. Wash the seeds 4–5 times with sterilized distilled water and culture on sterilized cotton moistened with sterile water.
3. Maintain the cultures at a temperature of 25 ± 2 °C, a 16-h photoperiod, and a light intensity of 50 μmol/m²/s with white fluorescent lights.
4. Use leaf explants with 15–30 mm in length and 20–25 mm in width from 45-day-old in vitro seedlings to initiate the cultures (*see* **Note 9**) (Fig. 3).

3.2 Agrobacterium
Culture Preparation

1. Prepare stock solution of 29.43 g acetosyringone in 1000 mL double-distilled water and store in –20 °C. First, add acetosyringone very slowly in 100 mL of DMSO. After complete dissolution, slowly add 900 mL distilled water into the dissolved acetosyringone solution.
2. Dissolve 4.27 g MES buffers in 1000 mL of distilled water and adjust its pH at 5.4. Sterilize the solution through 0.20 μM PTFE membrane. Store the solution at 4 °C.
3. Grow a single colony of *Agrobacterium rhizogenes* R1000 strain in LB medium (30 mL) at 28 °C for 12 h on a rotary shaker at 180 rpm in the dark.

Fig. 3 An experimental plant, *W. somnifera* habit. (**a**) Field-grown parent plant. (**b**) Collected seeds. (**c**) In vitro-raised plant for explant source. (**d**) Leaf explant for obtaining hairy roots

4. Pellet the bacterial cells by centrifugation at $7656 \times g$ for 10 min at 28 °C.

5. Wash the pellet twice with liquid half-strength MS basal medium (Murashige and Skoog 1962).

6. Dissolve the pelleted bacterial cells in the same medium (*see* **Note 10**).

7. Add acetosyringone (100 μM) and MES buffer to this bacterial suspension culture ($OD_{600} = 1$) 1 h before infection and keep this in rotary shaker at 200 rpm (*see* **Note 11**).

3.3 Hairy Root Induction from Leaf Explant

1. Prick (ten pricks/explant) the leaf explants with a sterile hypodermic needle on the midrib region of the lamina (*see* **Note 12**).

2. Immerse the wounded explants in the bacterial suspension culture for 15 min and blot on sterile tissue paper for 10 min (*see* **Note 13**).

3. Inoculate the dried explants in half-strength MS basal medium supplemented with 150 μM acetosyringone, MES buffer (20 mM), 3 % (w/v) sucrose, and 0.2 % (w/v) Phytagel.

4. Incubate the cultures in the dark at 25 ± 2 °C for 5 days.

5. Wash the explants first with sterilized distilled water followed by half-strength MS basal liquid medium, with 300 mg/L cefotaxime.

6. Transfer these explants into 30 mL half-strength MS basal medium with 3 % (w/v) sucrose, 0.2 % (w/v) Phytagel, and 300 mg/L cefotaxime.

7. Initiation of hairy roots could be observed at the wounded sites of the explants after about 8 days of culture (Fig. 4a).

8. Excise these hairy roots (2–3 cm in length) from the explants and transfer into 30 mL half-strength MS basal medium supplemented with 3 % (w/v) sucrose, 0.2 % (w/v) Phytagel, and 300 mg/L cefotaxime (*see* **Note 14**) (Fig. 4b–d).

9. Withdraw the antibiotic from the medium after the residual bacteria is completely killed and eliminated.

10. Maintain all the cultures under total darkness at 25 ± 2 °C.

3.4 Hairy Root Proliferation and Establishment

1. Inoculate 5 g FW of well-developed hairy roots (actively growing with 1 cm root length from root tip) in 250 mL Erlenmeyer flask containing 50 mL of half-strength MS basal

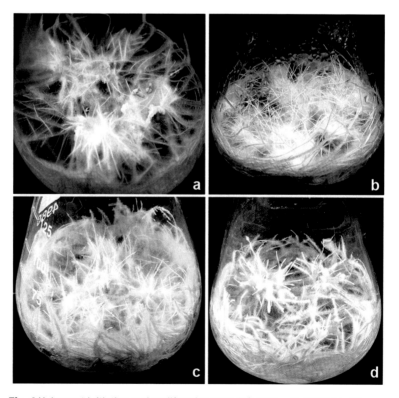

Fig. 4 Hairy root initiation and proliferation upon infection with R1000 in *W. somnifera* (**a–d**)

liquid medium with 3 % (w/v) sucrose (Fig. 5a) and keep the culture on a rotator shaker (80 rpm) at 25 ± 2 °C in the dark (*see* **Note 14**).

2. Inoculate 200 g (FW) of actively growing root tips (1 cm length from root tip) in 2000 mL of half-strength MS basal liquid medium containing 3 % (w/v) sucrose in 3 L Erlenmeyer flask (Fig. 5b) on a rotator shaker (120 rpm) at 25 ± 2 °C in the dark (*see* **Note 15**).

3. Harvest the hairy roots after 40 days of culture for the estimation of biomass and withanolide production (Table 3).

4. The hairy roots are subcultured four times at the interval of 7 days after the culture entered into log phase (14–40 days).

3.5 Assessment of Elicitation Conditions on Biomass Accumulation and Withanolide Production

1. Prepare stock solutions in 99.99 % ethanol for MeJ and in Milli-Q water for SA.

2. Dissolve 1.1 g of MeJ in 500 µL of 99.99 % ethanol and, after complete dissolution, add 500 µL of Milli-Q water to make up final stock. Likewise, dissolve 1 g of SA in 500 µL Milli-Q water and, after complete dissolution, add 500 µL of Milli-Q water to make up final stock. Filter the stocks via 0.20 µm

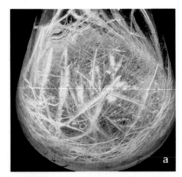

Fig. 5 Mass proliferation and biomass accumulation of *W. somnifera* hairy root culture in 250 mL flask (**a**) and 3 L flask (**b**)

Table 3

Effects of elicitation with MeJ and SA and exposure time on biomass accumulation in hairy root culture of *W. somnifera*. MeJ and SA were added to 30-day-old cultures

Elicitors (µM)	Exposure time (h)									
	0		2		4		6		8	
	FW	DW	FW	DW	FW	DW	FW	DW	FW	DW
MeJ										
0	26.36 ± 0.13^a	4.47 ± 0.11^a	26.36 ± 0.14^d	4.47 ± 0.12^b	26.36 ± 0.10^d	4.47 ± 0.18^a	26.36 ± 0.15^b	4.47 ± 0.13^c	26.36 ± 0.14^b	4.47 ± 0.19^b
5	26.36 ± 0.17^a	4.47 ± 0.16^a	24.41 ± 0.17^e	4.13 ± 0.14^e	18.62 ± 0.13^e	3.15 ± 0.15^a	13.73 ± 0.18^c	2.32 ± 0.17^f	7.26 ± 0.18^e	1.23 ± 0.14^e
10	26.36 ± 0.11^a	4.47 ± 0.14^a	20.32 ± 0.13^g	3.44 ± 0.13^g	14.43 ± 0.15^f	2.44 ± 0.13^a	9.68 ± 0.11^f	1.64 ± 0.11^g	4.43 ± 0.12^f	0.73 ± 0.17^f
15	26.36 ± 0.10^a	4.47 ± 0.10^a	14.52 ± 0.11^h	2.46 ± 0.17^h	8.28 ± 0.12^g	1.40 ± 0.10^a	3.31 ± 0.14^g	0.56 ± 0.15^h	1.72 ± 0.15^g	0.29 ± 0.16^f
20	26.36 ± 0.15^a	4.47 ± 0.17^a	8.82 ± 0.10^i	1.49 ± 0.19^i	3.22 ± 0.11^h	0.54 ± 0.11^a	1.12 ± 0.19^h	0.18 ± 0.13^i	0.43 ± 0.13^h	0.07 ± 0.13^g
SA										
0	26.36 ± 0.11^a	4.47 ± 0.15^a	26.36 ± 0.13^d	4.47 ± 0.16^c	26.36 ± 0.19^d	4.47 ± 0.16^a	26.36 ± 0.12^b	4.47 ± 0.15^c	26.36 ± 0.17^b	4.47 ± 0.14^b
50	26.36 ± 0.14^a	4.47 ± 0.13^a	22.35 ± 0.15^f	3.79 ± 0.14^f	26.74 ± 0.17^d	4.53 ± 0.13^a	20.66 ± 0.14^d	3.50 ± 0.13^e	18.55 ± 0.12^d	3.14 ± 0.16^d
100	26.36 ± 0.15^a	4.47 ± 0.11^a	25.83 ± 0.17^d	4.38 ± 0.10^a	29.48 ± 0.11^c	4.99 ± 0.15^a	24.43 ± 0.19^c	4.14 ± 0.12^d	22.34 ± 0.15^a	3.78 ± 0.18^d
150	26.36 ± 0.17^a	4.47 ± 0.19^a	29.56 ± 0.11^a	5.01 ± 0.09^a	32.68 ± 0.13^a	5.54 ± 0.17^a	28.12 ± 0.17^a	4.76 ± 0.14^a	27.61 ± 0.18^a	4.68 ± 0.13^a
200	26.36 ± 0.19^a	4.47 ± 0.10^a	27.77 ± 0.20^b	4.40 ± 0.18^b	30.72 ± 0.10^b	5.20 ± 0.10^a	26.68 ± 0.10^b	4.52 ± 0.10^b	24.12 ± 0.14^c	4.09 ± 0.11^c

The samples were taken at day 10 after elicitation (after 40 days of culture). Data represents the mean ± standard error of three independent experiments, each of which had three replicates. Mean values followed by the same letters within a column are not significantly different according to Duncan's multiple range test at 5 % level of probability

PTFE filters and store in brown bottles at –20 °C until further use. Before use in elicitation, thaw and mix well.

3. Use commercially available pre-sterilized ready-made filters for effective filtration.

4. Add MeJ at 15 μM and SA at 150 μM separately in the hairy root culture medium on 30-day-old cultures.

5. Add 99.9 % ethanol and Milli-Q water as controls in the hairy root culture medium on 30-day-old cultures (*see* **Note 16**).

6. Allow the hairy roots to grow in the presence of MeJ and SA separately for 4 h on 30-day-old cultures.

7. Re-culture the hairy roots again on fresh half-strength MS basal liquid medium without elicitors and harvest after 40 days for the estimation of biomass and withanolide yield (*see* **Note 16**).

8. Incubate the culture in 250 mL Erlenmeyer flask on a rotator shaker (80 rpm) at 25 ± 2 °C in darkness.

3.6 Estimation of Biomass Accumulation

1. Separate the hairy roots from the medium.

2. Rinse the hairy roots with sterile water and blot in the tissue paper to remove surface water.

3. Weigh the fresh weight (FW) of hairy roots using weighing balance.

4. Dry the roots constantly at 60 °C in oven for 2 days and record the dry weight (DW) (*see* **Note 17**).

3.7 Withanolide Extraction and HPLC Analysis

1. Grind the dried hairy roots into fine powder (1 g DW) with mortar and pestle using liquid nitrogen (*see* **Note 18**).

2. Extract with methanol (10 mL) under sonication for 30 min (*see* **Note 19**).

3. Keep the sample at 60 °C for 1 week in a vacuum desiccator.

4. Redissolve the remaining residue in 5 mL methanol and centrifuge at $7656 \times g$ for 15 min.

5. Filter the supernatant through 0.20 μm membrane for HPLC analysis.

6. Quantify withanolides by HPLC (Fig. 6).

7. Inject 20 μL of each syringe-filtered (0.20 μm) samples into the column and elute isocratically with methanol:water (65:35 v/v) at a flow rate of 1.0 mL/min.

8. Detect the withanolides with PDA detector set at a wavelength of 250 nm.

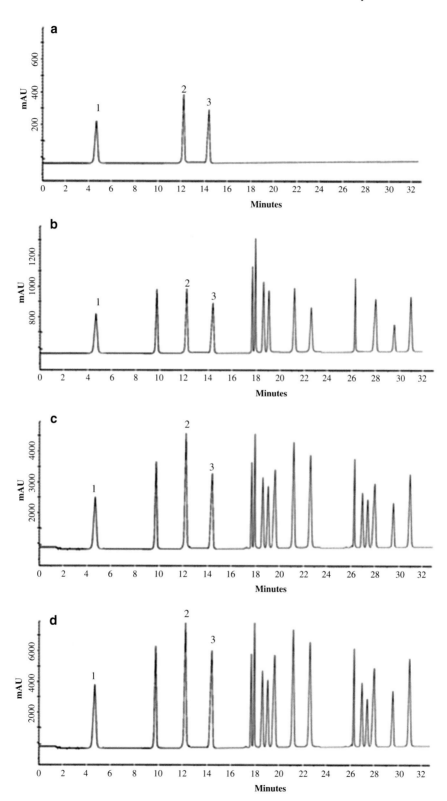

Fig. 6 HPLC chromatograms of withanolides in hairy root culture of *W. somnifera*. (**a**) Standards of withanolides (1-withanone, 2-withanolide A, and 3-withaferin A). (**b**) Control sample. (**c**) MeJ elicitation at 15 μM and 4-h exposure time. (**d**) SA elicitation at 150 μM and 4-h exposure time

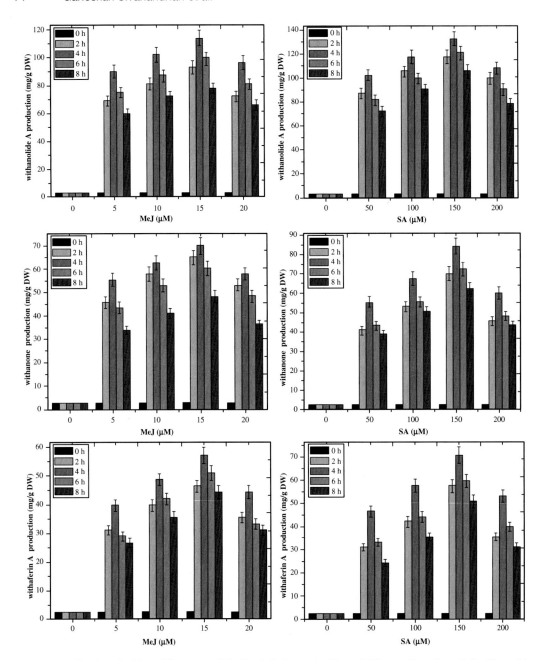

Fig. 7 Quantification of withanolides upon elicitation in hairy root culture of *W. somnifera*. (**a** and **d**) withanolide A, (**b** and **e**) withanone, and (**c** and **f**) withaferin A. The data were recorded after 40 days of culture. The cultures were established with 5 g FW of initial inoculum mass in 50 mL MS basal liquid medium

9. Calculate the relative amounts of withanolides by comparing their peak areas with standard curves generated using different amounts of external standards (Fig. 7).

10. Withanolide data are expressed as milligram per gram dry weight (Fig. 7).

11. Each sample should run in triplicate manner to check the consistency over the previous results.

4   Notes

1. There are many different varieties of *Withania somnifera* cultivated or naturally grown in India. We collected seeds from Kolli hills region of Southern India and cultivated them in our department experimental garden, Bharathidasan University. The transformation efficiency, hairy root induction, biomass, and withanolide production may vary with the genotype and habitat of the plant used.

2. The transformation efficiency, hairy root induction, biomass, and withanolide production may differ according to the *A. rhizogenes* strain used. We found that R1000 is the best strain for *Withania somnifera*.

3. A simple way of preparing plant culture medium starts by making up stock solutions of major inorganic nutrients, minor inorganic elements, iron source, vitamins, and individual plant growth regulators. The composition of the MS medium used here is shown in Table 2. Store the vitamins at –20 °C in small bottles. Before using this stock solution for medium preparation, defrost, and blend fully. The remaining MS stock solutions should be stored at 4 °C in the refrigerator, regularly checked, and removed if precipitation or sedimentation arises. Inorganic stock solutions should not be stored for longer than 1 month. It is much more desirable to prepare small stocks of growth regulators fresh for every batch of medium preparation, as little modification in stock concentration owing to precipitation can totally affect the tissue growth and development. Stock solutions of macro- and microelements and vitamins are also commercially available.

4. All solvents should be of HPLC grade and it should be filtered through 0.45 μM polytetrafluoroethylene (PTFE) filters before use. The solvents to be used for HPLC should be sonicated for better results.

5. The water should be replaced with cool water to avoid overheating during sonication.

6. Hot air oven should be maintained for complete dryness of the plant tissues at 60 °C for 2–3 days. The time may vary depending on the tissues used for drying. After complete dryness, the samples should be stored at –70 °C with tight sealing, and covered with parafilm and polyethylene bags until further use.

7. Make a stock solution of each withanolide standard by weighing 1 mg in aluminum foil and dissolve in 1 mL HPLC-grade methanol. Store this stock solution in an amber bottle at 4 °C for fresh standards. Transfer the required volume of the stock solution to room temperature before the experiment.

8. Squeeze the fully ripened fruits to take out the seeds. The seeds are fully washed with deionized water for the removal of unwanted debris. If the unwanted debris is on the surface of the seeds, it may affect the germination of seeds and lead to contamination. After clear wash, the seeds should be dried in muslin cloth under sunlight for 4–5 days. Store the seeds in tightly closed bottle at room temperature. The seeds germinated in the moistened cotton are prone to fungal or bacterial contamination. Hence, surface sterilization is done by using 0.1 % $HgCl_2$ rinse for 2 min; if the time exceeds beyond the optimum level, germination of the seeds will be reduced. Then, wash seeds thoroughly with sterilized distilled water and culture them in the moistened cotton. They should germinate rapidly. The inoculated seeds should be kept under darkness for 7 days in 16/8-h photoperiod.

9. Forty-five-day-old seedlings should show fully expanded leaf lamina with prominent vein system. The veins play an essential role in infection by bacteria and initiation of hairy root development. Appropriate trimming around the leaf lamina should be made to induce efficient bacterial infection.

10. The bacterial cell density should be at 0.1 OD_{600}.

11. Infectivity of *Agrobacterium* could be increased by adding acetosyringone (150 µM) and MES buffer (20 mM) to *Agrobacterium* cell suspension an hour before infection.

12. Care should be taken while pricking in the midrib region of the leaf. Gentle pricking is enough for wounding. If pricking exceeds ten times, *Agrobacterium* leaching and explant death will occur.

13. Immersion or infection time may differ for different explants and *Agrobacterium rhizogenes* strains. Exceeding the optimum time of infection will lead to explant death because of overgrowth of bacteria.

14. While transferring the induced root from the parent explant, care should be taken. Damaging the hairy root tip while transferring will reduce the proliferation, active growth, and biomass production.

15. Inoculum mass can play a vital role in biomass development and secondary metabolite production in most of the adventitious/hairy root cultures of several plant species.

16. Introduction of elicitor at the right stage of growing culture is crucial for secondary metabolite production. Before introducing the elicitor, standardization is required to determine at which stage of the root culture the elicitors should be added. Every plant species may differ in the growing pattern and the

growth curve. Hence the stage may be different for different species.

17. Hairy roots should be dried at 60 °C in a forced-air oven with 15 % moisture content for 2 days. After drying, the hairy root mass should be stored in screw-cap brown bottles at –20 °C and keep away from light and humidity until further use.

18. A proper fine grinding is necessary for successful extraction of secondary metabolites. Use liquid nitrogen for grinding of the plant samples.

19. While sonicating the plant extracts, heat is produced and it may affect the extraction step. Replace the bath reservoir with cool water every 5–8 min.

Acknowledgments

Prof. A. Ganapathi is thankful to the University Grants Commissions (UGC), Government of India, for the award of UGC-BSR (Basic Scientific Research Fellowship). The first author thankfully acknowledges the Council of Scientific and Industrial Research (CSIR), Government of India, for the award of CSIR-SRF. All authors acknowledge the CSIR, Government of India, for providing financial support to carry out this work. All the authors are grateful to Dr. S. Girija of the Department of Biotechnology, Bharathiar University, Coimbatore, for providing the HPLC facility.

References

1. Sivanandhan G, Kapil Dev G, Jeyaraj M, Rajesh M, Arjunan A, Muthuselvam M, Manickavasagam M, Selvaraj N, Ganapathi A (2013) Increased production of withanolide A, withanone and withaferin A in hairy root cultures of *Withania somnifera* (L.) Dunal elicited with methyl jasmonate and salicylic acid. Plant Cell Tiss Org Cult 114:121–129

2. Sivanandhan G, Arun M, Mayavan S, Rajesh M, Mariashibu TS, Manickavasagam M, Selvaraj N, Ganapathi A (2012) Chitosan enhances withanolides production in adventitious root cultures of *Withania somnifera* (L.) Dunal. Ind Crop Prod 37:124–129

3. Sangwan NS, Sangwan RS (2014) Secondary metabolites of traditional medicinal plants: a case study of Ashwagandha (*Withania somnifera*). In: Nick N, Opatrny Z (eds) Applied plant cell biology. Springer Berlin Heidelberg, Berlin, pp 325–367

4. Sivanandhan G, Selvaraj N, Ganapathi A, Manickavasagam M (2014) Enhanced biosynthesis of withanolides by elicitation and precursor feeding in cell suspension culture of *Withania somnifera* (L.) Dunal in shake-flask culture and bioreactor. PLoS One 9:1–11

5. Bhattacharaya SK, Bhattacharya D, Sairam K, Ghosal S (2002) Effect of *Withania somnifera* glycowithanolides on rat model of tardive dyskinesia. Phytomedicine 9:167–170

6. Sivanandhan G, Rajesh M, Arun M, Jeyaraj M, Kapil Dev G, Arjunan A, Manickavasagam M, Muthuselvam M, Selvaraj N, Ganapathi A (2013) Effect of culture conditions, cytokinins, methyl jasmonate and salicylic acid on the biomass accumulation and production of withano-

lides in multiple shoot culture of *Withania somnifera* (L.) Dunal using liquid culture. Acta Physiol Plant 35:715–728

7. Sivanandhan G, Arun M, Mayavan S, Rajesh M, Jeyaraj M, Kapil Dev G, Manickavasagam M, Selvaraj N, Ganapathi A (2012) Optimization of elicitation conditions with methyl jasmonate and salicylic acid to improve the productivity of withanolides in the adventitious root culture of *Withania somnifera* (L.) Dunal. Appl Biochem Biotechnol 168: 681–696

8. Sharada M, Ahuja A, Vij SP (2008) Applications of biotechnology in Indian ginseng (ashwagandha): progress and prospects. In: Kumar A, Sopory SK (eds) Recent advances in plant biotechnology and its applications. I. K. International Pvt Ltd, New Delhi, pp 645–667

9. Sivanandhan G, Kapil Dev G, Jeyaraj M, Rajesh M, Muthuselvam M, Selvaraj N, Manickavasagam M, Ganapathi A (2013) A promising approach on biomass accumulation and withanolides production in cell suspension culture of *Withania somnifera* (L.) Dunal. Protoplasma 250:885–898

10. Rao SR, Ravishankar GA (2002) Plant cell cultures: chemical factories of secondary metabolites. Biotechnol Adv 20:101–153

11. Pawar PK, Maheshwari VL (2004) *Agrobacterium rhizogenes*-mediated hairy root induction in two medicinally important members of family *Solanaceae*. Indian J Biotechnol 3:414–417

12. Kumar V, Kotamballi N, Chidambara M, Bhamid S, Sudha CG, Ravishankar GA (2005) Genetically modified hairy roots of *Withania somnifera* Dunal: a potent source of rejuvenating principles. Rejuvenation Res 8:37–45

13. Bandyopadhyay M, Jha S, Tepfer D (2007) Changes in morphological phenotypes and withanolide composition of Ri-transformed roots of *Withania somnifera*. Plant Cell Rep 26:599–609

14. Murthy HN, Dijkstra C, Anthony P, White DA, Davey MR, Power JB, Hahn EJ, Paek KY (2008) Establishment of *Withania somnifera* hairy root cultures for the production of withanolide A. J Int Plant Biol 50:975–981

15. Saravanakumar A, Aslam A, Shajahan A (2012) Development and optimization of hairy root culture systems in *Withania somnifera* (L.) Dunal for withaferin A production. Afr J Biotechnol 11:16412–16420

16. Sivanandhan G, Rajesh M, Arun M, Jeyaraj M, Kapil Dev G, Manickavasagam M, Selvaraj N, Ganapathi A (2012) Optimization of carbon source for hairy root growth and withaferin A and withanone production in *Withania somnifera*. Nat Prod Commun 7:1271–1272

17. Radman R, Saez T, Bucke C, Keshavarz T (2003) Elicitation of plants and microbial cell systems. Biotechnol Appl Biochem 37:91–102

18. Ketchum REB, Gibson DM, Croteau RB, Schuler ML (1999) The kinetics of taxoid accumulation in cell suspension cultures of *Taxus* following elicitation with methyl jasmonate. Biotechnol Bioeng 62:97–105

19. Wasternack C (2007) Jasmonates: an update on biosynthesis, signal transduction and action in plant stress response, growth and development. Ann Bot 100:681–697

20. Bonfill M, Mangas S, Moyano E, Cusido RM, Palazon J (2011) Production of centellosides and phytosterols in cell suspension cultures of *Centella asiatica*. Plant Cell Tiss Org Cult 104:61–67

21. Hu YH, Yu YT, Piao CH, Liu JM, Yu HS (2011) Methyl jasmonate- and salicylic acid-induced d-chiro-inositol production in suspension cultures of buckwheat (*Fagopyrum esculentum*). Plant Cell Tiss Org Cult 106: 419–424

22. Qu JG, Zhang W, Yu XJ (2011) A combination of elicitation and precursor feeding leads to increased anthocyanin synthesis in cell suspension cultures of *Vitis vinifera*. Plant Cell Tiss Org Cult 107:261–269

23. Kim OT, Bang KH, Kim YC, Hyun DY, Kim MY, Cha SW (2009) Upregulation of ginsenoside and gene expression related to triterpene biosynthesis in ginseng hairy root cultures elicited by methyl jasmonate. Plant Cell Tiss Org Cult 98:25–33

24. Rao MV, Lee H, Creelman RA, Mullet JE, Davis KR (2000) Jasmonic acid signalling modulates ozone-induced hypersensitive cell death. Plant Cell 12:1633–1646

25. Kang S, Jung H, Kang Y, Yun D, Bahk J, Yang J, Choi M (2004) Effects of methyl jasmonate and salicylic acid on the production of tropane alkaloids and the expression of PMT and H6H in adventitious root cultures of *Scopolia parviflora*. Plant Sci 166:745–751

26. Murashige T, Skoog F (1962) A revised medium for rapid growth and bioassays with tobacco tissue cultures. Physiol Plant 15: 473–497

Chapter 2

Stimulant Paste Preparation and Bark Streak Tapping Technique for Pine Oleoresin Extraction

Thanise Nogueira Füller*, Júlio César de Lima*, Fernanda de Costa, Kelly C.S. Rodrigues-Corrêa, and Arthur G. Fett-Neto

Abstract

Tapping technique comprises the extraction of pine oleoresin, a non-wood forest product consisting of a complex mixture of mono, sesqui, and diterpenes biosynthesized and exuded as a defense response to wounding. Oleoresin is used to produce gum rosin, turpentine, and their multiple derivatives. Oleoresin yield and quality are objects of interest in pine tree biotechnology, both in terms of environmental and genetic control. Monitoring these parameters in individual trees grown in the field provides a means to examine the control of terpene production in resin canals, as well as the identification of genetic-based differences in resinosis. A typical method of tapping involves the removal of bark and application of a chemical stimulant on the wounded area. Here we describe the methods for preparing the resin-stimulant paste with different adjuvants, as well as the bark streaking process in adult pine trees.

Key words Oleoresin, Pine, Tapping, Chemical stimulant, Wounding

1 Introduction

Several conifer species produce oleoresin, a complex mixture of isoprenoid compounds relevant for defense against herbivores and pathogens. Two major fractions can be recognized in oleoresin: (a) turpentine, the volatile fraction containing mono- and sesquiterpenes, and (b) rosin, the nonvolatile diterpene fraction. Oleoresin is a forest commodity of global interest, finding applications in diverse industry sectors. Rosin is used in adhesives, printing ink manufacture, and paper sizing. Turpentine can be used either as a solvent for paints and varnishes, or as a raw material for fractionation of high-value chemicals used in the pharmaceutical, agrochemical, and food industry [1–3].

During the extraction activity, resin is obtained from the tree in a similar way as rubber tree tapping, which generally involves the

---

*These authors have equally contributed to this work.

Arthur Germano Fett-Neto (ed.), *Biotechnology of Plant Secondary Metabolism: Methods and Protocols*, Methods in Molecular Biology, vol. 1405, DOI 10.1007/978-1-4939-3393-8_2, © Springer Science+Business Media New York 2016

preparation of the face of the tree, the installation of the resin collection system, tree wounding to induce resin flow, application of a chemical formulation to stimulate and maintain resin flow, collection of the resin, re-wounding of the tree, and application of the stimulant at suitable intervals [4].

Mechanical wound by itself induces defense responses in injured plants, but resin production can be chemically induced and modulated by applying stimulating pastes. Over the last 10 years, research has been conducted aiming at optimizing the production of induced oleoresin by the modification of the commercially used paste, which is basically composed of sulfuric acid or paraquat, N,N'-dimethyl-4,4$'$-bipyridinium dichloride (both involved in generating reactive oxygen species, resulting in increased and prolonged resin release by maximizing the effect of the wounding at the injury zone), and 2-chloroethylphosphonic acid (CEPA, an ethylene precursor, signaling molecule involved in stress-induced responses) [5, 6], which are the main active components that elicit the defense response. Although jasmonate is regarded as one of the best resin stimulators because of its capacity to regulate the expression of several genes encoding enzymes of high metabolic flux control in secondary metabolite biosynthesis [7–9], its use is restricted in large-scale operations due to high costs [1]. Several chemical adjuvants have been experimentally validated in the field to improve resin yield and/or to replace the need for the relatively expensive CEPA. Among these are included salicylic acid (mediator in plant responses to pathogens) [10] and auxin (stimulates both ethylene production and the differentiation of resin ducts) [11]. Alternative strategies to modify the composition of oleoresin stimulant paste are based on improving the activity of terpenoid synthases with metal cofactors and activators (e.g., K, Fe, and Mn) or by increasing ethylene sensitivity (adding Cu, a component of ethylene receptor proteins) [1, 3].

Oleoresin yield and quality are objects of interest in pine tree biotechnology, both in terms of environmental and genetic control. Monitoring these parameters in individual trees grown in the field provides a means to examine the control of terpene production in resin canals, as well as the identification of genetic-based differences in resinosis, providing a needed basis for generating super-resinous forests. A detailed protocol on how to prepare different types of stimulant pastes, as well as the basic procedure to tap pine trees using the bark streak technique, is described below.

2 Materials

2.1 Materials for Laboratory Preparation of Resin-Stimulating Pastes

1. Scale.

2. Beakers (500 mL capacity).

3. Graduated cylinder (100 mL capacity).

4. Glass rod.

5. Magnetic stirrer plate and magnetic rod.

6. Plastic squeeze bottles (600 mL capacity).

2.2   Materials for Pine Tapping in the Field

1. Plastic bags.

2. Wire (for belting the plastic bag on the tree).

3. Bark streaking tool or similar device to remove the bark.

4. Protective equipment (goggles, acid-resistant gloves, plastic apron, rubber boots).

5. Metal funnel for resin transfer from filled bags to storage containers.

6. Nylon or metal sieve for removal of debris from resin.

2.3   Reagents and Solutions

1. 98 % (v/v) sulfuric acid.

2. Naphthalene acetic acid.

3. Benzoic acid.

4. Potassium sulfate.

5. CEPA (Ethrel liquid plant growth regulator 240 g/L, Bayer, Calgary, AB, Canada) (*see* **Note 1**).

6. 1M sodium hydroxide (weigh 4 g of sodium hydroxide and prepare a solution of 100 mL with water).

7. Ground rice husk (1 kg of powder).

3   Methods

The four different chemical adjuvants were chosen according to their different modes of action (Table 1).

Usually the squeeze bottle used to apply paste supports about 400–500 mL. All methods herein described consider the total volume of 500 mL. With this volume you can treat approximately 400 trees. It is recommended to prepare fresh batches of resin-stimulating paste for each streaking period of a pine stand (i.e., once every 15 days) to avoid loss of activity by long storage periods.

3.1   Paste Preparation

1. For potassium paste, weigh 43.564 g of potassium sulfate for a 500 mM concentration solution. In a beaker, complete to a volume of 400 mL with water. Add 100 mL of sulfuric acid.

2. With the magnetic stirrer on, add 10 g of sifted rice husk (*see* **Note 2**) at a time until it reaches 200 g in total. Check if the paste sticks to a glass rod without dripping. This is the consistency for efficient stimulatory effect on the fresh wound cuts of the trees.

3. Add the prepared paste into the squeeze bottle (Fig. 1).

Table 1

Mode of action and responses expected for different adjuvants of resin-stimulating pastes evaluated in pine tapping operations (rationale of use)

Elicitor	Mode of action and responses observed
Sulfuric acid	Potential generator of free radicals; increases and prolongs gum resin yields by maximizing the effect of the wounding at the injury zone
Potassium	Activator of conifer terpenoid synthases
Salicylic acid	Mediator of plant responses to pathogens
Benzoic acid	Precursor of salicylic acid
Naphthalene acetic acid	Promotes ethylene biosynthesis (triggers 1-aminocyclopropane-1-carboxylic-acid—ACC—synthase gene transcription) and may induce resin duct formation, as a result of cambial activity stimulation
Copper	Component of ethylene receptors
CEPA	Precursor of ethylene, signaling molecule involved in stress-induced responses

Fig. 1 Squeeze bottles containing different stimulant pastes

4. For naphthalene acetic acid paste, weigh 0.093 g of naphthalene acetic acid for a 1 mM concentration solution. Add a few droplets of 1 M NaOH for helping the dissolution. In a beaker, complete to a volume of 400 mL with water. Add 100 mL of sulfuric acid.

5. With the magnetic stirrer on, add 10 g of sifted rice husk (*see* **Note 2**) at a time until it reaches 200 g in total. Check if the paste sticks to a glass rod without dripping.

6. Add the prepared paste into the squeeze bottle.

7. For benzoic acid paste, weigh 0.610 g of benzoic acid for a 10 mM concentration solution. Add a few droplets of 1 M NaOH for helping the dissolution. In a beaker, complete to a volume of 400 mL with water. Add 100 mL of sulfuric acid.

8. With the magnetic stirrer on, add 10 g of sifted rice husk (*see* **Note 2**) at a time until it reaches 200 g in total. Check if the paste sticks to a glass rod without dripping.

9. Add the prepared paste into the squeeze bottle.

10. For CEPA paste preparation, in a graduated cylinder, add 62.5 mL of Ethrel for a 3 % final concentration solution. Transfer it to a beaker. Complete to a volume of 400 mL of water. Add 100 mL of sulfuric acid.

11. With the magnetic stirrer on, add 10 g of rice husk sifted (*see* **Note 2**) at a time until it reaches 200 g in total. Check if the paste sticks to a glass rod without dripping.

12. Add the prepared paste into the squeeze bottle.

3.2   Tree Tapping

1. In the forest, select the trees for tapping (*see* **Note 3**).

2. In order to facilitate installation of the plastic bag for the first time, you should remove the rough outer bark from the area at the base of the tree, where the plastic bag will be fixed.

3. Belt the plastic bags with a wire around the trunk right above the area you intend tapping to collect the exuded oleoresin (Fig. 2).

4. With a bark shaving tool (Fig. 3), remove a horizontal strip of bark (with circa 2–3 cm in height) across about one-third of the width of the tree, above the plastic bag gutter, to promote resin flow. After bark removal, a clear whitish surface is exposed (cambium and younger sapwood).

5. Apply the stimulant paste on the top portion of the fresh groove formed by bark removal.

6. Remove the bark making horizontal grooves in the tree wood using a bark shaving tool or equivalent instrument every 15 days. The length of the grooves should be kept at about one-third of the tree circumference and the height should remain from 2 to 3 cm.

7. When the plastic bag is full, replace it and fix the new bag as close as possible of the newest groove (Fig. 2).

3.3   Initial Resin Harvest in the Forest

1. Collect the oleoresin accumulated in the bags.

2. If needed, carefully decant any rain water (it will form a separate phase from resin).

3. With a metal funnel, filter oleoresin through nylon or metal sieve to remove debris.

Fig. 2 Aspect of a tapped *Pinus elliottii* tree. The wounded area was formed by periodic streaking of the bark, followed by stimulant paste application. Plastic bags are belted around the trunk to collect the exuded oleoresin

4. Use the filtrate for the desired application (e.g., evaluation of oleoresin mass, terpene composition analyses by GC-MS and/or HPLC-MS, biotransformation reactions, distillation or further fractionation to purify higher value components).

4   Notes

1. CEPA is also known by different commercial names, such as Ethrel and Ethephon.

2. Usually, rice husk comes as large grains or pellets. Rice husk should be sifted through a relatively fine mesh filter to prevent obstruction of the squeeze bottle.

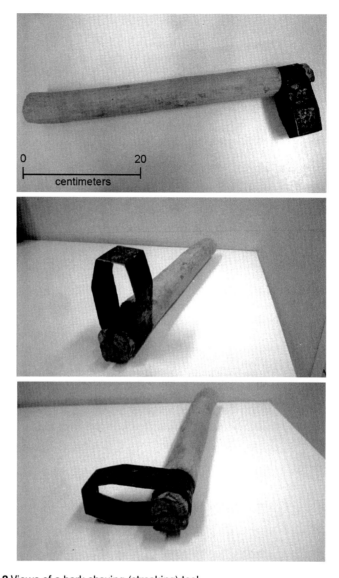

Fig. 3 Views of a bark shaving (streaking) tool

3. To begin, find a mature, live, good-sized, tight-barked pine tree for best results. Choose 10–20-year-old plantations or trees with diameters greater than 20–25 cm to tap. Pine species most suitable for tapping are *Pinus massoniana*, *P. yunnanensis*, *P. elliottii*, *P. caribaea*, *P. merkusii*, *P. kesiya*, *P. oocarpa*, *P. pinaster*, *P. roxburghii*, and *P. tabuliformis*. In most climates, oleoresin flow will be more pronounced in the spring and summer.

References

1. Rodrigues-Corrêa KCS, Fett-Neto AG (2012) Physiological control of pine resin production. In: Fett-Neto AG, Rodrigues-Corrêa KCS (eds) Pine resin: biology, chemistry and applications. Research Signpost, Kerala, p 25

2. Rodrigues-Corrêa KCS, de Lima JC, Fett-Neto AG (2012) Pine oleoresin: tapping green chemicals, biofuels, food protection, and carbon sequestration from multipurpose trees. Food Energy Secur 1:81–93

3. Rodrigues KCS, Apel MA, Henriques AT, Fett-Neto AG (2011) Efficient oleoresin biomass production in pines using low cost metal containing stimulant paste. Biomass Bioenergy 3:4442–4448

4. Coppen J, Hone GA (1995) Gum naval stores: naval stores: turpentine and rosin from pine resin. Non-wood forest products, vol 2. FAO, Rome, 63

5. Kossuth SV, Koch P (1989) Paraquat and CEPA stimulation of oleoresin production in lodgepole pine central stump-root system. Wood Fiber Sci 21:263–273

6. Wolter KE, Zinkel DF (1984) Observations on the physiological oleoresin synthesis in *Pinus resinosa*. Can J For Res 14:452–458

7. De Geyter N, Gholami A, Goormachtig S, Goossens A (2012) Transcriptional machineries in jasmonate-elicited plant secondary metabolism. Trends Plant Sci 17:349–359

8. Hudgins JW, Christiansen E, Franceschi VR (2004) Induction of anatomically based defense responses in stems of diverse conifers by methyl jasmonate: a phylogenetic perspective. Tree Physiol 24:251–264

9. Hudgins JW, Franceschi VR (2004) Methyl jasmonate-induced ethylene production is responsible for conifer phloem defense responses and reprogramming of stem cambial zone for traumatic resin duct formation. Plant Physiol 135:1–16

10. Rodrigues KCS, Fett-Neto AG (2009) Oleoresin yield of *Pinus elliottii* in a subtropical climate: seasonal variation and effect of auxin and salicylic acid based stimulant paste. Ind Crop Prod 30:316–320

11. Rodrigues KCS, Azevedo PC, Sobreiro LE, Pelissari P, Fett-Neto AG (2008) Oleoresin yield of *Pinus elliottii* plantations in a subtropical climate: effect of tree diameter, wound shape and concentration of active adjuvants in resin stimulating paste. Ind Crop Prod 27:322–327

Chapter 3

A Modified Protocol for High-Quality RNA Extraction from Oleoresin-Producing Adult Pines

Júlio César de Lima*, Thanise Nogueira Füller*, Fernanda de Costa, Kelly C.S. Rodrigues-Corrêa, and Arthur G. Fett-Neto

Abstract

RNA extraction resulting in good yields and quality is a fundamental step for the analyses of transcriptomes through high-throughput sequencing technologies, microarray, and also northern blots, RT-PCR, and RTqPCR. Even though many specific protocols designed for plants with high content of secondary metabolites have been developed, these are often expensive, time consuming, and not suitable for a wide range of tissues. Here we present a modification of the method previously described using the commercially available Concert™ Plant RNA Reagent (Invitrogen) buffer for field-grown adult pine trees with high oleoresin content.

Key words RNA, Pines, Concert plant RNA reagent, Stem RNA extraction, Oleoresin, Conifers

1 Introduction

Several conifer species, especially within the Pinaceae, have tissues with high concentrations of phenolics, terpenes, and polysaccharides [1]. Many specific protocols that are appropriate for plants rich in secondary metabolites have been developed, but the extraction of high-quality RNA from these tissues using commercial kits is often difficult and usually not applicable to woody tissues [2–6]. One of the major issues during RNA extraction concerns the presence of phenolic compounds, which oxidize and form quinones. Aromatic compounds bind RNA, which interferes in downstream steps and applications [3, 7]. Another point of concern is the harvest of plant samples in the experimental field, which constitutes another obstacle in the efforts to avoid degradation of RNA [8]. These problems often result in RNAs of low quality and insufficient amounts, especially for methodologies that normally require

---

*These authors have equally contributed to this work.

Arthur Germano Fett-Neto (ed.), *Biotechnology of Plant Secondary Metabolism: Methods and Protocols*, Methods in Molecular Biology, vol. 1405, DOI 10.1007/978-1-4939-3393-8_3, © Springer Science+Business Media New York 2016

more than hundreds of nanograms of nucleic acids to generate high-quality data, such as high-throughput sequencing.

Isolation of high-quality RNA is a crucial first step in gene expression and other genomic analyses. Regarding adult trees of conifer species, this task requires rigorous tissue collection procedures in the field and the employment of an RNA isolation protocol comprised of multiple organic extraction steps in order to isolate RNA of sufficient quality and quantity [1] (Figs. 1 and 2).

Efforts to overcome these limitations include the development of specific extraction methods, preferably without a high degree of complexity and at low cost. Considering all the statements and issues above, this chapter presents a reliable, relatively simple, and efficient modified method of RNA extraction from cambium and recent xylem of slash pine trees (*Pinus elliottii*) using Concert™ Plant RNA Reagent. The modifications proposed allow the investigator to routinely obtain high-quality RNA of tissues involved in oleoresin production in adult pine trees grown in the field.

2   Materials

2.1   Solutions and Buffer

1. 100 mL 2.5 M NaCl: Weigh 14.6 g of NaCl and make up to 100 mL with ultrapure water (prepared by purifying distilled and deionized water to attain a resistivity of 18 MΩ cm at 25 °C), autoclave at 121 °C and 1 atm for 20 min, and store at room temperature.

2. 100 mL 75 % ethanol: Add 75 mL of ethanol p.a. to a 100 mL graduated cylinder. Bring volume to 100 mL with ultrapure water. Cover with a plastic paraffin film (e.g., Parafilm) or equivalent and mix thoroughly by inversion.

3. 100 mL Chloroform p.a.

4. Concert™ Plant RNA Reagent (Life Science Biotechnology, Hamburg, Germany)—Lysis buffer: Must be kept at 4 °C.

5. 100 mL Isopropyl alcohol p.a.

6. Ultrapure water: Autoclave 500–1000 mL of ultrapure water for all preparations (121 °C and 1 atm for 20 min).

7. Liquid nitrogen (*see* **Note 1**).

8. Dry ice (*see* **Note 2**).

2.2   Materials for Collecting Tissue in the Field

1. Wood chisel for carving the bark.

2. Hammer for helping bark removal (*see* **Note 3**).

3. Gloves: Powder-free latex gloves should be used constantly.

4. Scraper to remove the RNA source tissue.

5. 15 mL RNase-free centrifuge tubes to collect pine tissue before RNA extraction.

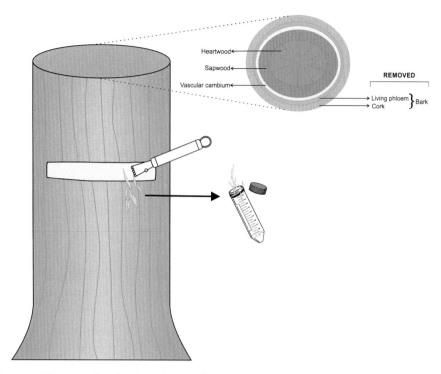

Fig. 1 Diagram of the procedure for tissue harvest for RNA extraction (cambium and recent sapwood)

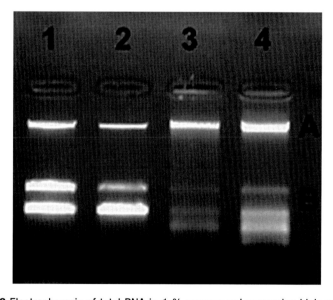

Fig. 2 Electrophoresis of total RNA in 1 % agarose gel comparing high-quality RNA and degraded RNAs. *Lane 1*: RNA of high quality; *lanes 2, 3*, and *4*: RNAs with a degradation pattern because of the several bands in the gel (especially patterns 3 and 4). (**a**) Contaminant DNA in the samples. (**b**) RNAs of ribosomal subunits. These RNAs must have two very sharp and clear bands (RNA pattern 1)

**2.3 Materials
for RNA Extraction**

1. Mortar and pestle.

2. 1.5 mL RNase-free microtubes: It is advisable to check their quality before use.

3. Refrigerated microcentrifuge: It must be at 4 °C before the start of the RNA extraction (*see* **Note 4**).

3 Methods

Before starting, RNase-free tips, tubes, glassware, and ultrapure water must be carefully prepared (autoclaving at 121 °C at 1 atm for 20 min is advisable).

1. In the field, add liquid nitrogen to a 15 mL centrifuge tube to pre-chill it.

2. Remove the bark using a wood chisel and a hammer. A whitish surface with the cambium should be exposed on the trunk (Fig. 1).

3. Scrape the exposed surface of the stem (cambium and recent xylem) using a scraper (for example, a lemon peel scraper) until you get enough tissue for RNA analysis (usually when the tissue covers 2 mL of the centrifuge tube). After removing the tissue, put it immediately into a centrifuge tube containing liquid nitrogen (*see* **Note 5**).

4. Keep the tubes frozen in liquid nitrogen until the addition of the lysis buffer.

5. At the laboratory, fill empty 1.5 mL microtubes with 500 μL pre-chilled lysis buffer (4 °C) (*see* **Note 6**).

6. Macerate the tissue in liquid nitrogen using mortar and pestle until you get a homogeneous fine powder. Add approximately 500 mg ground tissue to the microtube containing the chilled lysis buffer (*see* **Note 7**) and vortex until the ground tissue is fully resuspended (*see* **Note 2**).

7. Incubate the tube for 5 min at room temperature.

8. Centrifuge the samples at 12,000×*g* for 5 min at 4 °C.

9. Transfer the supernatant (450 μL) to an RNase-free 1.5 mL microtube (*see* **Note 8**).

10. Add 250 μL of 2.5 M NaCl and tap tubes to mix.

11. Add 400 μL of chloroform and mix thoroughly by inversion.

12. Centrifuge samples for 10 min at 12,000×*g* at 4 °C to separate the phases.

13. Transfer approximately 400 μL of the top aqueous phase to an RNase-free tube (*see* **Note 9**).

14. Add an equal volume of isopropyl alcohol, then mix, and let stand at room temperature for 10 min.

15. Centrifuge the sample at $12,000 \times g$ for 10 min at 4 °C.

16. Discard the supernatant carefully, to avoid losing the pellet (*see* **Note 10**).

17. Add 1 mL of cold 75 % ethanol to the pellet.

18. Centrifuge the sample at room temperature for 3 min at $12,000 \times g$.

19. Discard the supernatant carefully, to avoid losing the pellet.

20. Briefly, centrifuge the microtubes at $12,000 \times g$ for 1 min to collect the residual liquid and carefully remove it with a micropipette.

21. Dissolve the RNA pellet using 30 μL of ultrapure water with the help of a micropipette (*see* **Note 11**).

22. Store samples at –80 °C until use (*see* **Notes 2, 12,** and **13**). Prior to use, samples should be quantified and analyzed for integrity and quality (Figs. 2 and 3, Table 1).

4 Notes

1. In case of samples harvested in the field, liquid nitrogen must be stored adequately until the arrival in the experimental area. Care must be taken to have enough volume of liquid nitrogen to freeze all samples immediately after collection from the trees.

2. If the field is far from the laboratory, samples should be kept on dry ice until storage at –80 °C and/or the start of the RNA extraction procedure.

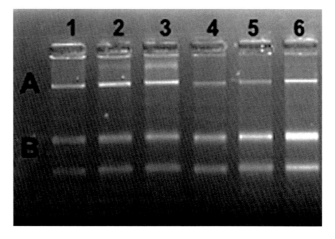

Fig. 3 Electrophoresis of total RNA in 1 % agarose gel. *Lanes 1, 2, 3, 4, 5,* and, *6*: Different volumes containing RNAs of high quality were applied in the gel. (**a**) Contaminant DNA in the samples. (**b**) RNAs of ribosomal subunits

Table 1
Quantification of pine RNA samples using Nanodrop equipment

Day of harvest[a]	Sample ID	ng/μL	ng of RNA amount in 30 μL
D0	1	126.4	3,792
	2	44.5	1,335
	3	112.6	3,378
	4	51.8	1,554
	5	134.8	4,044
D8	1	196.7	5,901
	2	236.8	7,104
	3	483.4	14,502
	4	137.7	4,131
	5	50.4	1,512

[a]Samples in this example were harvested in the experimental field at different days after mechanical stimulation for oleoresin production (D0: day 0; D8: day 8)

3. Usually, there are differences in the hardness of trees in the biological experiment. Care should be taken to remove the bark from those trees without damaging the inner tissue, which is the RNA source tissue.

4. The amount of solutions/reagents required depends on the number of samples in the biological experiment. It is always good to have solutions in a bit of excess.

5. It is very important that the tissue remains frozen at all the times to avoid oxidation and RNA degradation.

6. Concert™ RNA Reagent contains phenol. Manipulate the reagent in a fume hood. Wear gloves and eye protection when handling the reagent and solutions/steps containing it.

7. After the addition of the tissue to the buffer it can be stored at −80 °C for later extraction. It is crucial that the macerated tissue is in full contact with the buffer; otherwise thawing potentially results in RNA degradation.

8. Care must be taken not to collect/disturb the bottom pellet. If less cell debris is collected, lower amounts of contaminants will have to be discarded in the next steps.

9. It is important not to disturb the organic bottom phase during the collection of the top phase and transference.

10. The pellets must be white, not transparent or yellow to dark brown (this kind of pellet may indicate an oxidation or contamination coming from previous steps, e.g., the organic phase—**step 13**). In the last case, a re-extraction procedure is indicated.

11. If the solution seems cloudy, centrifuge at room temperature for 1 min at $12,000 \times g$ and transfer the supernatant to a new 1.5 mL microtube. A larger ultrapure water volume (up to 50 μL)

can be added in case the pellet is about half the volume indicated in **step 21** (30 µL).

12. At –80 °C, RNA samples may be stored for a long time. However, depending on the source organism and extraction method used, it is known that each nucleic acid sample (in solution) has its specific characteristics. From our experience with RNA pine samples, they can be stored at –80 °C for at least several months.

13. All RNA samples extracted with this procedure are contaminated with DNA (Figs. 1 and 2). A DNase treatment must be performed before cDNA synthesis.

References

1. Lorenz WW, Yu Y-C, Dean JFD (2010) An improved method of RNA isolation from Loblolly Pine (*P. taeda* L.) and other conifer species. J Vis Exp 36:1751–1756

2. MacKenzie DJ, McLean MA, Mukerji S, Green M (1997) Improved RNA extraction from woody plants for the detection of viral pathogens by reverse transcription-polymerase chain reaction. Plant Dis 81:222–226

3. Ghawana S, Paul A, Kumar H, Kumar A, Singh H, Bhardwaj PK (2011) An RNA isolation system for plant tissues rich in secondary metabolites. BMC Res Notes 4:85

4. Chomczynski P, Sacchi N (1987) Single-step method of RNA isolation by acid guanidinium thiocyanate-phenol chloroform extraction. Anal Biochem 162:156–159

5. Ding LW, Sun QY, Wang ZY, Sun YB, Xu ZF (2008) Using silica particles to isolate total RNA from plant tissues recalcitrant to extraction in guanidine thiocyanate. Anal Biochem 374:426–428

6. Camacho-Villasana YM, Ochoa-Alejo N, Walling L, Bray EA (2002) An improved method for isolating RNA from dehydrated and non- dehydrated Chili pepper (*Capsicum annuum* L.) plant tissues. Pl Mol Biol Rep 20:407–414

7. Liao Z, Chen M, Guo L, Gong Y, Tang F, Sun X, Tang K (2004) Rapid isolation of high-quality total RNA from *Taxus* and *Ginkgo*. Prep Biochem Biotechnol 34:209–214

8. Loomis MD (1974) Overcoming problems of phenolics and quinones in the isolation of plant enzymes and organelles. Methods Enzymol 31:528–544

Chapter 4

Collection of Apoplastic Fluids from *Arabidopsis thaliana* Leaves

Svend Roesen Madsen, Hussam Hassan Nour-Eldin, and Barbara Ann Halkier

Abstract

The leaf apoplast comprises the extracellular continuum outside cell membranes. A broad range of processes take place in the apoplast, including intercellular signaling, metabolite transport, and plant-microbe interactions. To study these processes, it is essential to analyze the metabolite content in apoplastic fluids. Due to the fragile nature of leaf tissues, it is a challenge to obtain apoplastic fluids from leaves. Here, methods to collect apoplastic washing fluid and guttation fluid from *Arabidopsis thaliana* leaves are described.

Key words Apoplastic fluid, Guttation fluid, Vacuum infiltration, Glucosinolates

1 Introduction

The leaf is an organ that is able to function both as source and sink tissue in long-distance transport processes; in addition leaves orchestrate their own intra-leaf distribution patterns. These transport processes involve not only photosynthates, but also specialized metabolites such as defense compounds. The leaf symplast represents the intracellular continuum of the cytoplasm interconnected via plasmodesmata. The leaf apoplast comprises the extracellular continuum outside the plasma membrane that includes cell walls, interstitial spaces, and xylem. The apoplast accommodates a highly dynamic and complex environment distinct from the symplast [1].

Plants produce an immense array of specialized metabolites, some of which are found in the apoplast [2–5]. Many specialized metabolites function as part of plants' defense system, and the presence of such compounds in the apoplast represents an early line of defense against many microbial pathogens. As an example, glucosinolates—compounds which constitute a major part of the

Arthur Germano Fett-Neto (ed.), *Biotechnology of Plant Secondary Metabolism: Methods and Protocols*, Methods in Molecular Biology, vol. 1405, DOI 10.1007/978-1-4939-3393-8_4, © Springer Science+Business Media New York 2016

chemical defense system in the model plant *Arabidopsis thaliana* (*Arabidopsis*) [6]—have been associated with apoplastic defense against virulent species of the bacteria *Pseudomonas* [7].

In recent studies using glucosinolates in *Arabidopsis* as model system, we showed that transport processes are important for establishing the distribution pattern of defense compounds [8–10]. To investigate intra-leaf transport processes, methods to isolate apoplastic fluids are essential. Due to the small size and fragility of *Arabidopsis* leaves, a challenge in investigating its apoplastic contents is to actually obtain the apoplastic fluids. We have employed two different approaches to obtain leaf apoplastic fluids (apoplastic washing fluid and guttation fluid) in order to investigate *Arabidopsis* intra-leaf transport processes of glucosinolates [8]. Apoplastic fluid is collected by first loading leaves with a washing solution (e.g., deionized water) using vacuum infiltration and then collecting the washing fluid (now containing apoplastic fluid) via centrifugation. After collecting apoplastic fluids it is important to verify that the extract has not been contaminated by intracellular content (e.g., due to cell breakage). Guttation fluid, composed of xylem sap and leaf apoplast solution, is the excess water released through hydathodes—secretory pores located at leaf margins [11]. Guttation takes place when stomata are closed, for example, at night and when the soil has a higher water potential than roots.

When collecting apoplastic fluids it is essential to avoid contamination with intracellular contents. The presence of cytoplasmic contamination in the collected apoplastic washing and guttation fluids can be tested by assaying for the presence of chlorophyll and cytosolic enzyme activities, e.g., malate dehydrogenase and glucose-6-phosphate dehydrogenase [2, 12]. In our lab, we test for cytoplasmic contamination by analyzing for glucosinolate hydrolysis products, as these only occur if cellular structure is disrupted and myrosinase (glucosinolate hydrolysis enzyme) released. Here, methods to collect and quantify extracellular fluids, as well as a method to test for intracellular contamination, are described.

2 Materials

2.1 Apoplastic Washing Fluid Collection Using Vacuum Infiltration

1. Razor blades.
2. Beaker.
3. Vacuum desiccator.
4. Vacuum pump.
5. Paper tissues.
6. Parafilm.
7. Falcon tubes (15 and 50 mL).
8. Sharp knife.

9. Tape.

10. Centrifuge.

11. Pipette tips.

12. Triton X-100 (or equivalent).

13. 0.3 M mannitol (or equivalent).

14. *Arabidopsis* plants with expanded leaves.

2.2 Guttation Fluid Collection

1. 200 μL Pipette tips.

2. 1.5 mL Microcentrifuge tubes.

3   Methods

3.1 Apoplastic Fluid Collection

1. Separate *Arabidopsis* leaves of similar size from the rosette (*see* **Note 1**) with a razor blade and submerge these briefly in deionized water in a beaker (100 mL or larger) to remove any surface contaminants (*see* **Note 2**).

2. Place leaves in another beaker (100 mL), submerge (*see* **Note 3**) in deionized water (80 mL—washing solution), and apply vacuum to infiltrate the leaves with the washing solution (approx. −70 kPa) using a vacuum pump and a desiccator. During vacuum, air is pulled out of extracellular spaces, and these are refilled/washed with water during release of the vacuum (*see* **Notes 4** and **5**). Slowly release and reapply the vacuum successively to reach 100 % infiltration, which can be estimated by dark green coloration of the water-saturated tissue (approx. 3×1 min. vacuum followed by release) (*see* **Notes 6** and **7**). After vacuum infiltration, gently dry the surface of the leaves using tissue paper. Do not leave any water puddles on the leaves, as this will interfere with the amount of apoplastic fluid collected after centrifugation (see next steps).

3. Place the leaves side by side on a sheet of parafilm (be careful not to break leaves at this point and try to avoid wrinkles—the leaves should be as stretched out as possible). Fold the parafilm to create a leaf-parafilm sandwich where the leaves are stacked between layers of parafilm (Fig. 1a, b). Gently (but relatively tightly so that the leaves stay in the parafilm sandwich during centrifugation—see below) mount the parafilm-leaf sandwich with tape onto the outside of a cylinder (we cut a 15 mL Falcon tube using a sharp knife—length of the cylinder should be approx. 1 cm longer than the leaf-parafilm sandwich). The leaf tips should be pointing downwards (Fig. 1c) (*see* **Note 8**). Do not block the openings in the parafilm layers where the leaf tips are protruding, as this is where apoplastic fluid exits the leaves (Fig. 1b).

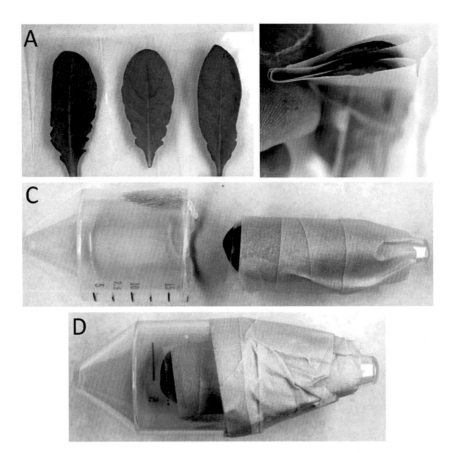

Fig. 1 Setup for collection of apoplastic washing fluid. (**A**) Example of three leaves to be stacked between layers of parafilm (one can also use separate pieces of parafilm). (**B**) Leaves stacked with leaf tips peeping out from the parafilm layers. (**C**) Leaves covered with parafilm are mounted onto a cut 15 mL Falcon tube with tape. (**D**) Secure the 15 mL Falcon tube mounted with leaves carefully and thoroughly with tape so it is "free-floating" inside the cut 50 mL Falcon tube (with leaf tips pointing towards the bottom of the 50 mL Falcon tube). After centrifugation, apoplastic washing fluid can be collected from the bottom of the cut 50 mL Falcon tube

4. Firmly mount the leaf-parafilm sandwich cylinder aggregate inside a cut 50 mL Falcon tube with tape (size of cut Falcon can be seen in Fig. 1c). The aggregate should be suspended so that the leaf tips do not touch the bottom of the 50 mL Falcon tube (do not refrain from using plenty of tape) (Fig. 1d). 3.6. Centrifuge in swinging buckets at 600–900 ×g for 10–15 min. The volume of apoplastic washing fluid (now at the bottom of the 50 mL Falcon tube) is measured with a pipette (**Notes 9** and **10**). After centrifugation leaves should be dry and not exhibit the dark green water-saturated color anymore.

5. Vacuum infiltration and centrifugation can be repeated to exhaust extraction of the desired apoplastic metabolite(s); however, the risk of cytoplasmic contamination will increase

with repetition due to the fragile nature of *Arabidopsis* leaves (*see* **Note 5**).

3.2 Guttation Fluid Collection

1. To stimulate guttation fluid production, place well-watered *Arabidopsis* plants at a desired developmental stage in high humidity overnight (*see* **Note 11**).

2. The next day, before onset of light in the growth chamber (to keep stomata closed), collect guttation fluid with a pipette (we used 200 μL tips) from the abaxial leaf side. Drops or sometimes small "guttation fluid puddles" are formed at hydathodes along leaf margins (*see* **Note 12**).

3.3 Measure of Glucosinolate Content and Analysis of Cytoplasmic Contamination in Collected Fluids

1. To test for cytoplasmic contamination, add 50 μL (the volume can be lower or higher than 50 μL, just note how much is added) water containing an exogenous glucosinolate (we used 0.02 mM sinigrin) to each collected apoplastic washing/guttation fluid sample followed by incubation for 5 min at room temperature to allow for any contaminating myrosinase activity to hydrolyze glucosinolates (*see* **Note 10**).

2. Analyze glucosinolate content in collected apoplastic washing or guttation fluids by high-pressure liquid chromatography (HPLC) or liquid chromatography mass spectrometry (LCMS) as described previously [8] (*see* **Note 13**) (see below for details on sample preparation and quantification of glucosinolates). Remember to include a control sample with only 50 μL water containing 0.02 mM sinigrin and no apoplastic fluid. Also include a control sample where leaves were intentionally damaged as a "positive control" for detection of intracellular enzymatic activity. Compare content of sinigrin in control samples to samples wherein apoplastic or guttation fluid has been collected. Sinigrin content equal to the control sample—wherein no apoplastic fluid had been collected—indicates no cytoplasmic contamination.

3. To measure glucosinolate content in apoplastic fluid and present data as nmol glucosinolate per μL apoplastic washing fluid, endogenous glucosinolates occurring in the apoplast may be normalized to an exogenous glucosinolate different from sinigrin: *p*-hydroxybenzyl glucosinolate (*p*OHB).

4. Incubate the collected apoplastic washing fluid (~35 μL) + 50 μL water containing 0.02 mM sinigrin for 5 min for the cytoplasmic contamination test (see above).

5. Add 100 μL 100 % methanol containing 0.02 mM *p*OHB to samples to stop the assay (aim for a final concentration of methanol in samples of at least 50 % (v/v) to minimize further glucosinolate hydrolysis).

6. Analyze endogenous glucosinolates present in the apoplast as well as the exogenous sinigrin by HPLC or LCMS and quantify by normalizing to *p*OHB (*see* **Note 13**) (*see* ref. 8).

7. This approach yields information about the concentration of metabolites in the collected volume of apoplastic washing fluid, but it does not give information about the metabolite concentration and volume of the "native" apoplastic fluid and air space present under normal conditions (i.e., leaves that are not vacuum infiltrated) (*see* **Note 14**). To obtain such data, it requires the co-infiltration of a standard compound in the infiltrated leaf, and is previously described (e.g., *see* ref. 2).

4   Notes

1. Larger leaves yield more apoplastic washing fluid and are easier to handle compared to smaller leaves. Therefore, if acceptable for the experiment, it is advantageous to grow plants under short-day conditions to obtain larger leaves. In our experiments, we use large source leaves from *Arabidopsis* plants before bolting, but we do not have experience with leaves and plants at other developmental stages. Please note that the leaf size influences the sizes of the plastic aggregates used later in the procedure. We collect leaves including the entire petiole, but the procedure should be applicable for leaves without petioles as well.

2. During the entire procedure, be very careful not to break the fragile *Arabidopsis* leaves, which will lead to cytoplasmic contamination of the apoplastic washing fluid.

3. To keep the leaves submerged, we pushed them down in the deionized water using a plate with holes (a sieve) that fit into the beaker. It is important that the plate does not seal the leaves in solution from the vacuum. That is why we used a plate with holes. Do not use a heavy plate that crushes the leaves against the beaker bottom, but one that just keeps them submerged.

4. Complete infiltration of leaf apoplastic spaces can be difficult to obtain. To aid infiltration, addition of Triton X-100 (or equivalent) to the infiltration fluid will help; however, be cautious as Triton X-100 is a detergent capable of lysing cell membranes, possibly leading to cytoplasmic contamination of the samples. Test out different concentrations and use the lowest volume of Triton X-100 that results in total infiltration of all leaves. In our experiments, 0.004 % (v/v) Triton X-100 gave the best results.

5. An alternative to water as infiltration fluid is a hypertonic 0.3 M mannitol solution. This solution will plasmolyze cells

and make them more resistant to breakage during repeated vacuum infiltrations [12].

6. Make sure that all leaves are submerged to obtain complete infiltration.

7. During vacuum, air bubbles can be seen on leaf surfaces. To maximize the extraction of air from leaf extracellular spaces and allow refilling with water when the vacuum is released, it is advisable to reduce the number of leaf surface air bubbles by gently bumping the desiccator containing the beaker with leaves against the lab bench during vacuum.

8. When stacking infiltrated leaves between layers of parafilm, there is room for ingenuity depending on the experimental setup and size of the leaves. We stack 1–5 leaves between layers of parafilm and mount this on the cut 15 mL Falcon tube (Fig. 1), but one can also make two parafilm stacks of leaves and mount these on either side of the Falcon tube. To protect the leaves make sure that they are completely separated from each other by parafilm, except from the very leaf tips from which the fluid is collected, and which may "peep out." Perform a couple of test experiments to find a suitable number of leaves in the parafilm sandwich and the optimal tightness of the tape when mounting the sandwich onto the Falcon tube.

9. From a single leaf that weighs approx. 150 mg, we normally obtain approx. 35 μL apoplastic washing fluid.

10. If the metabolite(s) of interest requires immediate stabilization in a solvent (e.g., methanol) different from the infiltration fluid, the apoplastic fluid can be centrifuged directly into the solvent by placing a suitable volume of it in the bottom of the cut 50 mL Falcon tube before centrifugation. However, the solvent may interfere with the cytoplasmic contamination assay (for example, most enzymes do not function in higher concentrations of methanol). A solution is to centrifuge at 4 °C if that helps to stabilize the compound(s) and immediately thereafter divide the apoplastic fluid sample into two and place in different tubes: To one tube add a solvent to stabilize the metabolite(s) of interest (e.g., methanol), and use the other tube to perform the cytoplasmic contamination assay.

11. A simple humidity chamber can be made by placing plants in an open box with a wet cloth in the bottom, and by closing the box with transparent plastic (e.g., Saran wrap).

12. Guttation fluid often yields a larger volume per leaf compared to apoplastic washing fluid collection. Furthermore, as guttation fluid can be collected from entire *Arabidopsis* rosettes, it is an easy way to collect from leaves at different developmental stages.

13. If necessary for detection of metabolites, samples can be concentrated before analysis via, e.g., lyophilization (freeze-drying) or centrifugal vacuum concentration.

14. To obtain information about metabolite concentration and volume of the "native" apoplastic fluid in a fast and easy manner, we tried to simply centrifuge "native" apoplastic fluid out of normal non-vacuum-infiltrated leaves, but this was unsuccessful. In an alternative approach, we weighed leaves' pre-vacuum infiltration and post-centrifugation in order to use the weight difference as a volume for "native" apoplastic fluid (assuming a density of 1 g/mL); however, this method did not give meaningful results, as the post-centrifugation weight was sometimes higher than the pre-vacuum infiltration weight. This suggests that not all apoplastic washing fluid is recovered during centrifugation, and that relative (not absolute) values are obtained.

References

1. Sattelmacher B (2001) Tansley review no. 22 - the apoplast and its significance for plant mineral nutrition. New Phytol 149:167–192
2. Baker CJ, Kovalskaya NY, Mock NM, Owens RA, Deahl KL, Whitaker BD, Roberts DP, Hammond RW, Aver'yanov AA (2012) An internal standard technique for improved quantitative analysis of apoplastic metabolites in tomato leaves. Physiol Mol Plant Pathol 78:31–37
3. Robinson JM, Bunce JA (2000) Influence of drought-induced water stress on soybean and spinach leaf ascorbate-dehydroascorbate level and redox status. Int J Plant Sci 161:271–279
4. Lee YW, Jin S, Sim WS, Nester EW (1995) Genetic evidence for direct sensing of phenolic compounds by the VirA protein of *Agrobacterium tumefaciens*. Proc Natl Acad Sci U S A 92:12245–12249
5. Foyer CH, Descourvières P, Kunert KJ (1994) Protection against oxygen radicals: an important defence mechanism studied in transgenic plants. Plant Cell Environ 17:507–523
6. Halkier BA, Gershenzon J (2006) Biology and biochemistry of glucosinolates. Annu Rev Plant Biol 57:303–333
7. Fan J, Crooks C, Creissen G, Hill L, Fairhurst S, Doerner P, Lamb C (2011) *Pseudomonas sax* genes overcome aliphatic isothiocyanate-mediated non-host resistance in *Arabidopsis*. Science 331:1185–1188
8. Madsen SR, Olsen CE, Nour-Eldin HH, Halkier BA (2014) Elucidating the role of transport processes in leaf glucosinolate distribution. Plant Physiol 166:1450–1462. doi:10.1104/pp. 114.246249
9. Andersen TG, Nour-Eldin HH, Fuller VL, Olsen CE, Burow M, Halkier BA (2013) Integration of biosynthesis and long-distance transport establish organ-specific glucosinolate profiles in vegetative *Arabidopsis*. Plant Cell 25:3133–3145
10. Nour-Eldin HH, Andersen TG, Burow M, Madsen SR, Jorgensen ME, Olsen CE, Dreyer I, Hedrich R, Geiger D, Halkier BA (2012) NRT/PTR transporters are essential for translocation of glucosinolate defence compounds to seeds. Nature 488:531–534
11. Pilot G, Stransky H, Bushey DF, Pratelli R, Ludewig U, Wingate VPM, Frommer WB (2004) Overexpression of glutamine dumper1 leads to hypersecretion of glutamine from hydathodes of *Arabidopsis* leaves. Plant Cell 16:1827–1840
12. Boudart G, Jamet E, Rossignol M, Lafitte C, Borderies G, Jauneau A, Esquerre-Tugaye MT, Pont-Lezica R (2005) Cell wall proteins in apoplastic fluids of *Arabidopsis thaliana* rosettes: identification by mass spectrometry and bioinformatics. Proteomics 5:212–221

Chapter 5

From Plant Extract to a cDNA Encoding a Glucosyltransferase Candidate: Proteomics and Transcriptomics as Tools to Help Elucidate Saponin Biosynthesis in *Centella asiatica*

Fernanda de Costa, Carla J.S. Barber, Darwin W. Reed, and Patrick S. Covello

Abstract

Centella asiatica (L.) Urban (Apiaceae), a small annual plant that grows in India, Sri Lanka, Malaysia, and other parts of Asia, is well-known as a medicinal herb with a long history of therapeutic uses. The bioactive compounds present in *C. asiatica* leaves include ursane-type triterpene sapogenins and saponins—asiatic acid, madecassic acid, asiaticoside, and madecassoside. Various bioactivities have been shown for these compounds, although most of the steps in the biosynthesis of triterpene saponins, including glycosylation, remain uncharacterized at the molecular level. This chapter describes an approach that integrates partial enzyme purification, proteomics methods, and transcriptomics, with the aim of reducing the number of cDNA candidates encoding for a glucosyltransferase involved in saponin biosynthesis and facilitating the elucidation of the pathway in this medicinal plant.

Key words *Centella asiatica*, Saponin, Glucosyltransferase, Proteomics, Transcriptomics

1 Introduction

The perennial subshrub *Centella asiatica* (L.) Urban (Apiaceae) has been widely cultivated in China, Southeast Asia, India, Sri Lanka, Africa, and Oceania. It has been used in the treatment of a wide variety of afflictions such as skin diseases, rheumatism, inflammation, syphilis, mental illness, epilepsy, hysteria, dehydration, and diarrhea [1]. Some of the medicinal properties of the species have been attributed to characteristic ursane-type saponins (namely madecassoside and asiaticoside), present predominantly in aerial parts of *C. asiatica* [2]. Despite the fact that *C. asiatica* saponins have been well studied due to their biological importance, the

Arthur Germano Fett-Neto (ed.), *Biotechnology of Plant Secondary Metabolism: Methods and Protocols*, Methods in Molecular Biology, vol. 1405, DOI 10.1007/978-1-4939-3393-8_5, © Springer Science+Business Media New York 2016

genes involved in the biosynthesis of the saponin backbone and subsequent backbone modifications are still relatively unknown.

Glycosylation is a key modification of plant natural products during their biosynthesis, enhancing solubility and stability and facilitating storage and accumulation in plant cells [3]. Glycosylation of natural products is catalyzed by uridine diphosphate-sugar-dependent glycosyltransferases (UGTs), enzymes that catalyze the transfer of sugar residues from uridine diphosphate to an acceptor, most typically an alcohol [4]. In spite of the medicinal importance of ursane saponins in *C. asiatica*, relatively little is known about the late stages of their biosynthesis, including the glycosylation steps [5].

Nowadays, the currently available next-generation sequencing (NGS) methods have dramatically increased the speed and ease with which sequence data can be obtained from RNA samples from a wide range of non-model species [6]. Furthermore, NGS has provided a rapid improvement in throughput, read length, and accuracy compared to other techniques [7]. As a consequence, NGS technologies have become the strategy of choice for the discovery of genes encoding enzymes responsible for the multistep formation of plant specialized metabolites, since the effectiveness of selecting candidate genes that encode enzymes with metabolic functions is dependent on the amount of information available about the catalytic steps in a particular biosynthetic pathway [8]. With the aid of bioinformatics tools, the identification of novel biosynthetic genes is an achievable goal.

In this chapter, we present an approach that integrates partial enzyme purification, proteomics methods, and transcriptomics, with the aim of reducing the number of cDNA candidates encoding enzymes that are involved in glycosylation of triterpene backbones. The elucidation of key genes involved in saponin production has great biotechnological relevance and the implementation of microbial engineering is an important step in the development of a versatile strategy with the possibility of commercial production.

2   Materials

Starting with a purified protein extract and with the help of proteomic methods and cDNA sequence data, a full-length cDNA clone encoding the first glucosyltransferase from *C. asiatica* was obtained and cloned in *Escherichia coli*.

Prepare all solutions using distilled water and analytical grade reagents. Prepare and store all reagents at 4 °C (unless indicated otherwise).

2.1 Protein Extract from Leaves of C. asiatica

1. Prepare 1 L of 100 mM Tris–HCl pH 7.6: Add about 800 mL of water to a 1 L flask or beaker. Weigh 12.1 g of Tris base and transfer to the cylinder. Add water to a volume of 900 mL. Mix and adjust pH to 7.6 with the appropriate volume of concentrated HCl. Weigh 27.32 g of sorbitol and add to the buffer solution, resulting in a final concentration of 150 mM in solution. Pipette 0.89 mL of β-mercaptoethanol, resulting in 12.5 mM in solution. Add 500 μL of plant protease inhibitor cocktail (Sigma Chemical Company, St. Louis, MO, USA). Add 1 mM of phenylmethanesulfonylfluoride (PMSF), a serine protease inhibitor (*see* **Note 1**).

2. Approximately 1 g of plant material (fresh weight).

3. 100 mg of polyvinylpolypyrrolidone (PVPP).

4. Disposable PD-10 Desalting Columns (GE HealthCare, Uppsala, Sweden).

2.2 Optimized Assay for Enzyme Activity in the Extract

1. Total reaction volume: 100 μL. In a microcentrifuge tube, add the following compounds: 250 μM UDP-glucose (or other donor substrate), 1 mM acceptor substrate—asiatic acid or madecassic acid, 20 mM bicine pH 8.5, 1 mM dithiothreitol, 10 mM MgCl$_2$, 0.2 % (w/v) of bovine serum albumin, 10 μL of the protein extract.

2. Reverse-phase column for LC/MS analysis: Zorbax 80A Extended-C18 column, 5 μm particle size, 2.1×150 mm (Agilent Technologies, Santa Clara, CA, USA).

3. Mobile-phase A: Water-acetonitrile (90:10 v/v) containing 0.1 % (v/v) formic acid and 0.1 % (w/v) ammonium formate and mobile-phase B: water-acetonitrile (10:90 v/v) containing 0.1 % (v/v) formic acid and 0.1 % (w/v) ammonium formate.

2.3 Anion-Exchange Chromatography

1. Mono-Q 5/50 GL column (GE HealthCare, Uppsala, Sweden).

2. Agilent 1100 series HPLC equipped with auto injector, a diode array detector, and a fraction collector maintained at 5 °C.

3. Mobile-phase A: 20 mM Tris pH 8.0 containing 2 mM 2-mercaptoethanol and mobile-phase B: 20 mM Tris pH 8.0 containing 2 mM 2-mercaptoethanol and 1 M NaCl.

2.4 SDS-PAGE of AEC Fractions

1. Precision plus protein standards (Bio-Rad, Hercules, CA, USA).

2. Polyacrylamide gels with a 10-well comb for 50 μL samples: 4–15 % Mini-PROTEAN TGX Gel (Bio-Rad, Hercules, CA, USA).

3. Running buffer: 10× Tris/glycine/SDS (Bio-Rad, Hercules, CA, USA).

4. Stain: Oriole fluorescent gel stain solution 1× (Bio-Rad, Hercules, CA, USA).

2.5 Quadrupole Time-of-Flight Liquid Chromatography Mass Spectrometry (Q-TOF LC/MS)

1. Q-TOF Global Ultima mass spectrometer (Micromass, Manchester, UK) equipped with a nano-electrospray (ESI) source and a nanoACQUITY UPLC solvent delivery system (Waters, Milford, MA).

2. Q-TOF parameter settings: Capillary voltage of 3850 V, a cone voltage of 120 V, and a source temperature of 80 °C.

3. Mobile phase C: 0.2 % formic acid and 3 % acetonitrile and mobile phase D: 0.2 % formic acid and 95 % acetonitrile.

2.6 Bioinformatics Analysis Using Software MASCOT

1. MASCOT MS/MS software ion search parameters: Maximum of one missed cleavage allowed for tryptic digestion, mass tolerance for precursor peptide ions of ±50 ppm and for fragment ions up to ±0.4 Da. Carbamidomethylation of cysteine was selected as a fixed modification, and oxidation of methionine was used as a variable modification.

3 Methods

3.1 Protein Extract from Leaves of C. asiatica

1. Collect 1 g of plant material and grind with PVPP 2 % using a mortar and pestle containing liquid N_2.

2. Put ground sample in a 15 mL centrifuge tube (on ice) containing 5 mL of buffer plus PMSF and mix using a vortexer for 5 min.

3. Centrifuge for 10 min at $12,000 \times g$ at 4 °C.

4. Transfer the supernatant recovered from centrifugation to a PD-10 column (2 mL in each column).

5. Add 25 mL buffer (without PMSF) to wash column and discard eluate.

6. Add 2 mL sample in each and discard eluate.

7. Add 0.5 mL buffer and discard eluate.

8. Add 3 mL buffer and keep the eluate (*see* **Note 2**).

3.2 Assay for Enzyme Activity in the Extract

1. Glucosyltransferase enzyme assays should be carried out in a total volume of 100 µL containing 10 µL of crude protein extract (1.8–2.2 mg/mL of protein), 250 µM UDP-Glc, 1 mM acceptor substrate, 20 mM Bicine, pH8.5, 1 mM DTT, 0.2 % BSA, and 10 mM $MgCl_2$.

2. Add enzyme to start reaction.

3. Incubate the reaction mixture at 30 °C for 1 h.

4. Stop reaction by transferring to –80 °C.

5. Analyze samples using ion trap LC/MS (below).

6. Analysis of enzyme reactions is performed using an Agilent 6320 ion trap LC/MS system with electrospray source under default Smart Parameter settings (scanning in the m/z range of 50–2200 at 8100 mass units/s with an expected peak width

of ≤0.35 mass units), with a reverse-phase column maintained at 35 °C.

7. The separation gradient is 90:10 A/B to 100 % B over 30 min, constant in 100 % B until 35 min, 100 % B to 90:10 A/B until 35.4 min, and constant 90:10 A/B rate until 50 min.

3.3 Anion-Exchange Chromatography

1. Submit 1 mL of crude protein extract to anion-exchange chromatography (AEC).

2. Pre-equilibrate column with 20 mM Tris–HCl, pH 8.0.

3. Perform elution with a mobile phase of 20 mM Tris–HCl, pH 8.0, containing a linear gradient from zero to 50:50 A:B for 50 min changed to 100 % B at 55 min and maintained at 1 M NaCl for 10 min. Collect 1 mL samples, assay, and analyze by LC/MS to detect enzyme activity. Submit active samples to SDS-PAGE. Non-active samples must also be tested and used as a negative control.

3.4 SDS-PAGE of AEC Fractions

1. Samples should be mixed 1:1 by volume with 2× SDS-PAGE loading sample buffer (200 mM Tris pH 6.8, 4 % (w/v) SDS, 0.2 % (v/v) bromophenol blue, 20 % (v/v) glycerol, 200 mM DTT) and heated at 95 °C for 5 min.

2. Run samples in SDS-PAGE under denaturing conditions using electrophoresis buffer (25 mM Tris–HCl, pH 7.5, 250 mM glycine, 0.1 % SDS w/v) and a 4–15 % (w/v) polyacrylamide gradient Ready Gel.

3. Precision Plus Protein™ Unstained Standard (Bio-Rad) must be loaded on the same gel.

4. Stain the gel with Oriole™ fluorescent gel stain (Bio-Rad) overnight.

5. Visualize the protein bands with UV light.

6. Excise bands of interest from active fractions of the SDS-PAGE gel and put each one in separate microcentrifuge tubes. The proteins are then automatically destained, reduced with DTT, alkylated with iodoacetamide, and digested with porcine trypsin (sequencing grade, Promega, Madison, WI, USA) using a MassPREP protein digest station and following recommended procedures (Micromass, Manchester, UK).

3.5 Q-TOF LC/MS

1. Trypsinized peptides are desalted with an in-line solid-phase trap column (180 μm × 20 mm) packed with 5 μm resin (Symmetry C18; Waters, Milford, MA) and separated on a capillary column (100 μm × 100 mm; Waters) packed with BEH130 C18 resin (1.7 μm; Waters Milford, MA) using a column temperature of 35 °C.

2. Five microliter samples are introduced into the trap column at a flow rate of 15 μL/min for 3 min, using C:D 99:1, and flow

is diverted to waste. After desalting, the flow is routed through the trap column to the analytical column with a linear gradient of 1–10 % solvent D (400 nL/min, 16 min), followed by a linear gradient of 10–45 % solvent D (400 nL/min, 30 min).

3.6 Bioinformatics Analysis Using MASCOT

1. The data are processed with ProteinLynx Global Server 2.4 (Waters) using RAW files from Q-TOF LC/MS. The resulting PKL files were analyzed using MASCOT (version 2.3.02; Matrix Science Ltd., London, UK) for peptide searches against the NCBI nr database (version 011110) hosted by the National Research Council of Canada and local databases containing the sequence information from Illumina and Roche 454 (this work) sequencing from methyl jasmonate (MeJA)-induced root cultures of *C. asiatica* cDNA [6].

2. Ten cDNA candidates were selected and subsequently cloned (not discussed in this chapter). One of them was the first glucosyltransferase found in *C. asiatica*, responsible for the addition of the first glucose on the backbone (UGT73AD1).

4   Notes

1. PMSF must be prepared as an anhydrous ethanol solution, since it is rapidly degraded in water. It must be added to the Tris buffer at the moment of its use. Care must be taken handling this compound. PMSF is toxic if swallowed and can cause severe skin burns and eye damage.

2. To regenerate PD-10 columns add 25 mL of extraction buffer and 25 mL of Tris pH 8.0. Keep the column closed at 4 °C.

References

1. Yu Q-L, Duan H-Q, Takaishi Y, Gao W-Y (2006) A novel triterpene from *Centella asiatica*. Molecules 11:661–665
2. Sangwan RS, Tripathi S, Singh J, Narnoliya LK, Sangwan NS (2013) *De novo* sequencing and assembly of *Centella asiatica* leaf transcriptome for mapping of structural, functional and regulatory genes with special reference to secondary metabolism. Gene 525:58–76
3. Wang X (2009) Structure, mechanism and engineering of plant natural product glycosyltransferases. FEBS Lett 583:3303–3309
4. Vogt T, Jones P (2000) Glycosyltransferases in plant natural product synthesis: characterization of a supergene family. Trends Plant Sci 5:380–386
5. Shibuya M, Nishimura K, Yasuyama N, Ebizuka Y (2010) Identification and characterization of glycosyltransferases involved in the biosynthesis

of soyasaponin I in *Glycine max*. FEBS Lett 584:2258–2264
6. Xiao M, Zhang Y, Chen X, Lee EJ, Barber CJ, Chakrabarty R, Desgagné-Penix I, Haslam TM, Kim YB, Liu E, MacNevin G, Masada-Atsumi S, Reed DW, Stout JM, Zerbe P, Zhang Y, Bohlmann J, Covello PS, De Luca V, Page JE, Ro DK, Martin VJ, Facchini PJ, Sensen CW (2013) Transcriptome analysis based on next-generation sequencing of non-model plants producing specialized metabolites of biotechnological interest. J Biotechnol 166:122–134
7. Zhang J, Chiodini R, Badr A, Zhang G (2011) The impact of next-generation sequencing on genomics. J Genet Genomics 38:95–109
8. Facchini PJ, Bohlmann J, Covello PS, De Luca V, Mahadevan R, Page JE, Ro DK, Sensen CW, Storms R, Martin VJ (2012) Synthetic biosystems for the production of high-value plant metabolites. Trends Biotechnol 30:127–131

Chapter 6

Production of Recombinant Caffeine Synthase from Guarana (*Paullinia cupana* var. *sorbilis*) in *Escherichia coli*

Flávia Camila Schimpl, Eduardo Kiyota, and Paulo Mazzafera

Abstract

Caffeine synthase (CS) is a methyltransferase responsible for the last two steps of the caffeine biosynthesis pathway in plants. CS is able to convert 7-methylxanthine to theobromine (3,7-dimethylxanthine) and theobromine to caffeine (1,3,7-trimethylxanthine) using *S*-adenosyl-L-methionine as the methyl donor in both reactions. The production of a recombinant protein is an important tool for the characterization of enzymes, particularly when the enzyme has affinity for different substrates. Guarana has the highest caffeine content among more than a hundred plant species that contain this alkaloid. Different from other plants, in which CS has a higher affinity for paraxanthine (1,7-dimethylxanthine), caffeine synthase from guarana (PcCS) has a higher affinity for theobromine. Here, we describe a method to produce a recombinant caffeine synthase from guarana in *Escherichia coli* and its purification by affinity chromatography. The recombinant protein retains activity and can be used in enzymatic assays and other biochemical characterization studies.

Key words Caffeine synthase, Guarana, *Paullinia cupana*, Recombinant expression, Alkaloid

1 Introduction

Caffeine is a purine alkaloid with important economic value and is present in different food products, such as coffee, tea, chocolate, cola drinks, beauty products, and medicines [1]. Among more than a hundred plant species, guarana (*Paullinia cupana*) contains the highest caffeine content described to date: 2.5–6.5 % of the dry weight of the seeds. In addition to being used in a widely consumed beverage with the same name, guarana is known for its stimulating and medicinal benefits [2].

Caffeine has been extensively studied in coffee and tea plants, both of which have very similar caffeine biosynthesis pathways. Recent reports have also shown that the main pathway is present in cocoa, mate, and guarana [3, 4]. The pathway starts with the for-

Arthur Germano Fett-Neto (ed.), *Biotechnology of Plant Secondary Metabolism: Methods and Protocols*, Methods in Molecular Biology, vol. 1405, DOI 10.1007/978-1-4939-3393-8_6, © Springer Science+Business Media New York 2016

mation of xanthosine by four possible routes from purine nucleotides and occurs in three methylation steps: xanthosine → 7-methylxanthosine → 7-methylxanthine → theobromine (3,7-dimethylxanthine) → caffeine (1,3,7-trimethylxanthine). S-adenosyl-L-methionine (SAM) is the methyl donor [3].

Caffeine synthase (CS, EC 2.1.1.160) is a bifunctional enzyme that acts in the last two steps of caffeine biosynthesis, converting 7-methylxanthine to theobromine and theobromine to caffeine [5]. Recombinant CS enzymes from coffee (CCS1–AB086414 and CaDMXT1–AB084125) and tea (TCS1–AB031280) have affinity for the substrate 7-methylxanthine and theobromine, but curiously, the highest affinity is for paraxanthine (1,7-dimethylxanthine), which has not been found to date in the tissues of caffeine-containing plants [5–7]. Different from other species, CS from guarana (PcCS–BK008796.1) has a high affinity for theobromine but also for 7-methylxanthine [4].

The heterologous production of proteins from eukaryotic organisms is a powerful tool for their characterization, both functionally and structurally. Thus, the production of recombinant PcCS guarana may enable various functional studies with the enzyme [4] and can also be used in additional experiments, such as crystallographic characterization, enhancing our knowledge of its structure and biological function. Here, we describe the experimental procedure to produce a recombinant caffeine synthase from guarana.

2 Materials

2.1 Cloning

1. Sodium perchlorate buffer, 5 M sodium perchlorate, 0.3 M Tris–HCl pH 8.3, 2 % (w/v) PEG 4000, 1 % (w/v) SDS, 8.5 % (w/v) PVPP, 1 % (v/v) β-mercaptoethanol.
2. Ethanol: Absolute and 70 %.
3. Phenol/chloroform/isoamyl alcohol (25:24:1).
4. 3 M Sodium acetate pH 5.2.
5. DNase "Turbo DNA-free" (Ambion, Inc.).
6. "SuperScript III First-Strand" (Invitrogen).
7. Specific primers: Forward 5'-GGATCCATGGATATGAAA GATGTGCTTTG-3' and reverse 5'-GTCGACTTAATTTC TCTTCAAACCAACAACA-3' (the underlined bases represent the restriction sites for the enzymes BamHI and SalI, respectively), 0.25 μM final concentration.
8. "Platinum® PCR SuperMix High Fidelity" (Invitrogen).
9. "GeneJET Gel Extraction Kit" (Fermentas).
10. "pGEM®-T easy Vector System" (Promega).

11. Competent *Escherichia coli* strain DH5α (Novagen).

12. Restriction enzyme *Eco*RI (Promega).

13. "PureLink Quick Plasmid Miniprep Kit" (Invitrogen).

14. "BigDye Terminator v3.1 Cycle Sequencing Kit" (Applied Biosystems).

15. Expression vector pET28a (Invitrogen).

16. Restriction enzymes *Bam*HI and *Sal*I (Fermentas).

17. Cloning *E. coli* strain DH10B (Invitrogen).

18. Expression strain *E. coli* BL21(pRIL) (Stratagene Agilent).

2.2 Protein Expression and Preparation of Bacterial Lysate

1. Luria-Bertani (LB) medium: Add 10 g tryptone, 5 g yeast extract, and 10 g NaCl in 950 mL deionized water and shake until the solutes have dissolved. Adjust the pH to 7.0 with 5 N NaOH (approx. 0.2 mL) and complete the volume to 1 L with deionized water. Sterilize by autoclaving for 20 min at 15 psi (1.05 kg/cm²) on the liquid cycle. For solid medium, add 15 g of agar before autoclaving. Store at 4 °C or at room temperature.

2. Dissolve kanamycin (Sigma) 50 mg/mL in deionized water, filter-sterilize, and store at –20 °C. Dissolve chloramphenicol (Sigma) 34 mg/mL in ethanol and store at –20 °C.

3. Isopropyl β-D-thiogalactoside (IPTG): Dissolve 2 g of IPTG (Sigma) in 10 mL distilled water to produce a 1 M stock solution and sterilize by filtration through a 0.22 μm disposable filter. Dispense the solution into 1 mL aliquots and store at –20 °C.

4. Affinity buffer: 50 mM Potassium phosphate, pH 7.3, 100 mM NaCl, 5 % glycerol.

5. Lysozyme: Dissolve lysozyme (Roche) 10 mg/mL in distilled water and store at –20 °C.

2.3 SDS-Polyacrylamide Gel Electrophoresis (SDS-PAGE)

1. Resolving buffer (4×): 1.5 M Tris–HCl, pH 8.8. Store at room temperature.

2. Stacking buffer (4×): 1.0 M Tris–HCl, pH 6.8. Store at room temperature.

3. 30 % Acrylamide/bis-acrylamide solution: Prepare 30 % (w/v) acrylamide and 0.8 % (w/v) bis-acrylamide in deionized water. Store in the dark at 4 °C.

4. SDS 10 %: Prepare 10 % (w/v) of sodium dodecyl sulfate in deionized water. Store at room temperature.

5. Ammonium persulfate: Prepare 10 % (w/v) ammonium persulfate in deionized water and immediately freeze in single-use aliquots at –20 °C.

6. Tetramethylethylenediamine (TEMED).

7. Isopropanol: Pure isopropanol.

8. Loading buffer (2×): 100 mM Tris–HCl, pH 6.8, 20 % glycerol, 2 % β-mercaptoethanol, 0.2 % (w/v) bromophenol blue, and 4 % SDS. Aliquot and store at room temperature. Dilute 1/1 with sample.

9. Running buffer (10×): 250 mM Tris, 1.92 M glycine, 1 % (w/v) SDS.

10. Molecular weight marker: "EZ-Run™" protein marker (Fisher Scientific).

11. Staining solution: 45 % (v/v) methanol, 10 % (v/v) acetic acid, 0.27 % (w/v) Coomassie Brilliant Blue R250 (Sigma) in distilled water. Mix for 2 h and filter through Whatman paper number 1. Store in a fume hood at room temperature.

12. Destaining solution: 10 % (v/v) methanol, 5 % (v/v) acetic acid in distilled water. Store in a fume hood at room temperature.

2.4 Purification by IMAC: Immobilized Metal Affinity Chromatography

1. Agarose resin: Ni-NTA agarose (Qiagen) (*see* **Note 1**). Stored in 20 % ethanol at 4 °C.

2. Elution buffer: 50 mM Potassium phosphate, pH 7.2, 100 mM NaCl, 5 % glycerol.

3. Imidazole: 1 M Stock solution in affinity buffer (*see* **Note 2**).

3 Methods

3.1 Selection of Caffeine Synthase Sequence

1. From a Guarana EST database (Rede da Amazônia Legal de Pesquisas Genômicas—Realgene) [8], we selected a caffeine synthase sequence that exhibited the highest similarity to the caffeine synthase genes from coffee (*Coffea arabica*: CCS1, AB086414, 1203 bp) and tea (*Camellia sinensis*, TCS1, AB031280, 1438 bp). Both sequences are available in GenBank.

2. The guarana CS contig, of 1369 bp (PcCS-BK008796), was used for primer design in the 5′UTR and 3′UTR regions. As we produced the recombinant PcCS to study its enzymatic activity, this protocol is based on the use of the recombinant protein. Although we believe that the quantity of the protein was quite reasonable, it might be that for other purposes, such as structural studies, the protocol may need to be adapted to improve the quantity of protein in the soluble fraction.

3.2 RNA Extraction

1. Total RNA is extracted from immature guarana seeds of cultivar BRS-Amazonas using a small-scaled modified protocol of the sodium perchlorate method [9]: in 2 mL microcentrifuge tubes, 100 mg of ground sample is homogenized with 1.5 mL

extraction buffer and kept shaking (160 rpm) for 10 min, followed by 5-min incubation on ice and centrifugation ($6000 \times g$, 4 °C, 10 min). The supernatant is recovered, 1 volume of phenol/chloroform/isoamyl alcohol (25:24:1) is added, and the tube is vortexed thoroughly to mix the phases. After centrifugation ($10,000 \times g$, 4 °C, 10 min) the aqueous phase (intermediary) is recovered, ethanol (2–2.5 volumes) is added, and the solution is gently but thoroughly mixed, and left resting for 20 min at –20 °C for RNA precipitation. After centrifugation ($10,000 \times g$, 4 °C, 10 min), the supernatant is discarded and the pellet carefully washed with 70 % ethanol, followed again by centrifugation ($10,000 \times g$, 4 °C, 5 min). The supernatant is discarded again and the pellet dried at room temperature by tapping the tube. The pellet is solubilized in 750 μL TE buffer (containing 5 μL 2-ME) and 750 μL of phenol/chloroform/isoamyl alcohol (25:24:1) is added. The mixture is vortexed for 2–3 min and centrifuged ($8000 \times g$, 4 °C, 10 min). The supernatant is collected and transferred to a new tube. The extraction is repeated with phenol/chloroform/isoamyl alcohol (25:24:1). The supernatants are then pooled and 0.1 volume of 3 M sodium acetate pH 5.2 and 2–2.5 volumes of ethanol are added and mixed by inverting the tube several times. After 1 h at –80 °C the mixture is centrifuged ($10,000 \times g$, 4 °C, 30 min), the liquid phase is discarded, and the pellet is washed with 70 % ethanol. The ethanol is eliminated by centrifugation ($10,000 \times g$, 4 °C, 5 min) and the pellet dried at room temperature by inverting the tube. The RNA is solubilized in 30 μL DEPC water. The RNA is then treated with DNase.

3.3 Caffeine Synthase Sequence Amplification and Cloning

1. Total RNA (2 μg) is used to produce the first-strand cDNA for further amplification of the full-length open reading frame by RT-PCR using specific primers, and the amplified fragment is purified from agarose gel using "GeneJET Gel Extraction Kit" (Fermentas). The complete coding sequence is subcloned into the pGEM®-T Easy vector (according to the manufacturer's recommendations) and transferred to DH5α competent cells (Novagen). To confirm the presence of the insert, the plasmid is extracted by plasmid mini-preparation "PureLink Quick Plasmid Miniprep Kit" (Invitrogen) and digested with EcoRI (according to the manufacturer's recommendations). The identity of the digestion product is obtained by sequencing with "BigDye Terminator v3.1 Cycle Sequencing Kit" (Applied Biosystems). The plasmid is then digested with BamHI and SalI, and the CS insert is cloned into an expression vector (pET28a, Novagen) (according to the manufacturer's recommendations). The expression vector is transferred primarily into E. coli DH10B and then in E. coli BL21(pRIL) by electroporation. Fusion tags can facilitate the detection and purification of

a target protein, and pET28a is able to produce fusion proteins carrying the His-tag at the N and/or C termini (*see* **Note 3**). In our case, we chose a His-tag at the N-terminus; therefore, we added a "stop" codon to the reverse specific primer.

3.4 Bacterial Growth and Protein Extraction

1. Pre-inoculate transformed *E. coli* in 5 mL LB containing 50 μg/mL kanamycin (5 μL stock solution diluted to 5 mL with LB medium) and 102 μg/mL chloramphenicol (15 μL stock solution diluted to 5 mL with LB medium) and incubate overnight in a rotatory shaker (200 rpm) at 37 °C.

2. Determine the $OD_{600\ nm}$ of the pre-inoculum and dilute it to $OD_{600\ nm} = 0.1$ with 500 mL LB containing 50 μg/mL kanamycin and 102 μg/mL chloramphenicol and grow at 37 °C in a rotatory shaker (200 rpm) until an $OD_{600\ nm} = 0.6–0.8$.

3. Reserve 1 mL aliquot as the "non-induced fraction" (control) and induce expression by adding IPTG (final concentration of 1 mM) to the remaining volume (*see* **Note 4**).

4. Incubate the culture at 37 °C with shaking (200 rpm) for 4 h (*see* **Note 5**).

5. Harvest the cells by centrifugation at $5000 \times g$ and 4 °C for 10 min.

6. Solubilize the cell pellet in 10 mL affinity buffer and keep the solute all times on ice.

7. Add lysozyme to 0.08 mg/mL final concentration and incubate on ice for 30 min.

8. To guarantee cell disruption, sonicate the lysate using a microtip sonicator, applying pulses of 3 s with 5 s intervals until reaching 3 min of sonication time. Keep the extract in an ice bath.

9. Centrifuge the lysate at $15,000 \times g$ and 4 °C for 20 min, recover the protein soluble phase (supernatant), and retain it for further purification. A sample of the insoluble phase (pellet) should be recovered and mixed 1/1 (w/v) with SDS-PAGE loading buffer to analyze by electrophoresis the amount of protein retained in this phase (Fig. 1, *see* also **Note 6**).

3.5 Purification by IMAC: Immobilized Metal Affinity Chromatography

1. Incubate the soluble phase with 1 mL of Ni-NTA agarose resin for 1 h at 4 °C with shaking (50 rpm).

2. Pack the mixture in a column for gravity purification (you can use an empty PD-10 Sephadex G25 column from GE-Healthcare or any similar column) with a bottom filter. Alternatively, an ordinary syringe can be prepared with glass wool. Collect the flow-through for SDS-PAGE analysis.

3. Elute the bound proteins with elution buffer supplemented with imidazole to a final concentration ranging from 10 to 500 mM. First, apply 1 mL of elution buffer containing 10 mM imidazole (the lowest imidazole concentration, prepared using

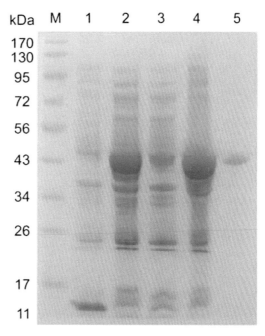

Fig. 1 Analysis of PcCS expression from the pET28a vector and purification of the recombinant protein by SDS-12 % PAGE stained with Coomassie Blue. *M* = molecular weight marker; *1* = crude extract of transformed *E. coli* without IPTG induction (control); *2* = *E. coli* crude extract after induction with IPTG for 4 h at 37 °C; *3* = induced soluble fraction; *4* = induced insoluble fraction; *5* = pooled fractions purified by affinity chromatography with Ni-NTA agarose resin. Reproduced from Schimpl et al. [4] with permission from Elsevier

the 1 M imidazole stock solution) onto the column, collect the flow-through, then add another 1 mL of the same solution, and collect the new flow-through separately. Repeat this two-step elution procedure with elution buffers of increasing imidazole concentration up to 500 mM (*see* **Note** 7). Keep all fractions on ice for further SDS-PAGE analysis.

3.6 SDS-PAGE Analysis of Purified Fractions

1. These instructions apply to the Mini protean III minigel system (BioRad) but are easily adaptable to other systems. Prepare a 12 % resolving gel by mixing 1.7 mL resolving buffer, 3 mL 30 % acrylamide mix, 75 μL 10 % SDS, 75 μL 10 % ammonium persulfate, 3 μL TEMED, and complete to 7.5 mL with deionized water. Pour the gel solution onto the Miniprotean casting gel apparatus (approximately 6 cm), leaving a space for a stacking gel, and overlay with isopropanol, according to the manufacturer's instructions. When the remaining gel solution is polymerized, the gel is also polymerized. Discard the isopropanol by inverting the gel casting apparatus and dry carefully the top of the gel with a piece of filter paper.

2. Prepare a 5 % stacking gel solution by mixing 380 μL stacking buffer, 500 μL 30 % acrylamide solution, 30 μL 10 % SDS, 30

μL 10 % ammonium persulfate, 3 μL TEMED, and complete to 7.5 mL with deionized water. Pour the gel solution onto the polymerized gel and insert the comb. When the remaining gel solution is polymerized, the gel is also polymerized. Pour the running buffer into the Mini protean casting gel apparatus and carefully remove the comb.

3. Prepare samples for SDS-PAGE analysis by mixing 10 μL of 2× sample buffer and 10 μL of each fraction. Boil the samples for 5 min in a water bath and let them cool at room temperature. Quickly centrifuge the samples in a benchtop centrifuge and load 10 μL of the samples into the wells. Reserve the external wells for the protein molecular weight marker.

4. Run the gel at 100 V until the samples migrate through the stacking gel and then apply 150 V for protein separation in the resolving gel until the migration blue dye reaches the bottom of the gel.

5. Tap out the gel from the apparatus and discard the stacking gel. Submerge the gel in a sufficient volume of staining solution and agitate slowly for 30 min.

6. Rinse the gel in distilled water and submerge in a sufficient volume of destaining solution. Agitate slowly for 30–60 min, and replace with new destaining solution, repeating periodically until the gel background is clear. Use caution, however, because excessive destaining may lead to the loss of band intensity. Store the gel in water at room temperature for further documentation.

4 Notes

1. Immobilized metal ion affinity chromatography (IMAC) is based on the interaction of metals with certain amino acids, especially histidine. Thus, proteins with affinity for certain metals can be separated from a pool of proteins. Nickel and copper, for instance, are used for purifying His-tagged proteins.

2. Mix different volumes of the 1 M imidazole stock solution with 2× affinity buffer to obtain different concentrations of imidazole. Typical solutions have 10, 30, 50, 100, 200, 300, and 500 mM imidazole.

3. In general, the His-Tag sequence is a very useful fusion partner for protein purification. It is especially useful for those proteins initially expressed as inclusion bodies because affinity purification can be accomplished under the high salt-denaturing conditions used to solubilize proteins.

4. The IPTG concentration may be changed to improve protein expression (0.1–1 mM).

5. Changes in the temperature and time of induction may improve the total amount of protein expressed in the soluble fraction.

We tested another induction temperature (20 °C), but it was not as good as the temperature indicated (37 °C).

6. The insoluble fraction contains more protein than the soluble fraction (Fig. 1). The overproduction of proteins can lead to the formation of inclusion bodies, as the capacity of the cellular machinery responsible for protein folding in the native form is compromised [10]. As denaturing conditions typically irreversibly destroy enzyme activity, we chose to use the soluble portion from the supernatant phase. Depending on the goal of the work, it may be possible to alter the temperature, IPTG concentration, and time of induction to increase the amount of protein in the soluble fraction.

7. SDS-PAGE shows the imidazole concentration fraction at which CS was eluted. Therefore, those fractions containing CS at the desired purity (least possible background) could be pooled to compose a working CS sample; we selected the fractions from 200 to 500 mM. Because a high imidazole concentration is required for protein purification, a dialysis procedure may be implemented to eliminate the salt content of the sample and proceed with functional studies.

References

1. Mazzafera P (2012) Which is the by-product: caffeine or decaf coffee? Food Energy Secur 1:70–75

2. Schimpl FC, da Silva JF, Gonçalves JFC, Mazzafera P (2013) Guarana: revisiting a highly caffeinated plant from the Amazon. J Ethnopharmacol 150:14–31

3. Ashihara H, Ogita S, Crozier A (2011) Purine alkaloid metabolism. In: Ashihara H, Crozier A, Komamine A (eds) Plant metabolism and biotechnology. Wiley, Chichester, pp 163–189

4. Schimpl FC, Kiyota E, Mayer JLS, Gonçalves JFC, da Silva JF, Mazzafera P (2014) Molecular and biochemical characterization of caffeine synthase and purine alkaloid concentration in guarana fruit. Phytochemistry 105:25–36

5. Kato M, Mizuno K, Crozier A, Fujimura T, Ashihara H (2000) Caffeine synthase gene from tea leaves. Nature 406:956–957

6. Mizuno K, Okuda A, Kato M, Yoneyama N, Tanaka H, Ashihara H, Fujimura T (2003) Isolation of a new dual-functional caffeine synthase gene encoding an enzyme for the conversion of 7-methylxanthine to caffeine from coffee (*Coffea arabica* L.). FEBS Lett 534:75–81

7. Uefuji H, Ogita S, Yamaguchi Y, Koizumi N, Sano H (2003) Molecular cloning and functional characterization of three distinct N-methyltransferases involved in the caffeine biosynthetic pathway in coffee plants. Plant Physiol 132:372–380

8. Ângelo PCS, Nunes-Silva CG, Brígido MM, Azevedo JSN, Assunção EN, Sousa ARB, Patrício FJB, Rego MM, Peixoto JCC, Oliveira WP Jr, Freitas DV, Almeida ERP, Viana AMHA, Souza AFPN, Andrade EV, Acosta POA, Batista JS, Walter MEMT, Leomil L, Anjos DAS, Coimbra RCM, Barbosa MHN, Honda E, Pereira SS, Silva A, Pereira JO, Silva ML, Marins M, Holanda FJ, Abreu RMM, Pando SC, Gonçalves JFC, Carvalho ML, Leal-Mesquita ERRBP, Da Silveira MA, Batista WC, Atroch AL, França SC, Porto JIR, Schneider MPC, Astolfi-Filho S (2008) Guarana (Paullinia cupana var. sorbilis), an anciently consumed stimulant from the Amazon rain forest: the seeded-fruit transcriptome. Plant Cell Rep 27:117–124

9. Rezaian MA, Krake LR (1987) Nucleic acid extraction and virus detection in grapevine. J Virol Methods 17:277–285

10. Patra AK, Mukhopadhyay R, Mukhija R, Krishnan A, Garg LC, Panda AK (2000) Optimization of inclusion body solubilization and renaturation of recombinant human growth hormone from *Escherichia coli*. Protein Expr Purif 18:182–192

Chapter 7

Heterologous Expression and Characterization of Mimosinase from *Leucaena leucocephala*

Vishal Singh Negi and Dulal Borthakur

Abstract

Heterologous expression of eukaryotic genes in bacterial system is an important method in synthetic biology to characterize proteins. It is a widely used method, which can be sometimes quite challenging, as a number of factors that act along the path of expression of a transgene to mRNA, and mRNA to protein, can potentially affect the expression of a transgene in a heterologous system. Here, we describe a method for successful cloning and expression of mimosinase-encoding gene from *Leucaena leucocephala* (leucaena) in *E. coli* as the heterologous host. Mimosinase is an important enzyme especially in the context of metabolic engineering of plant secondary metabolite as it catalyzes the degradation of mimosine, which is a toxic secondary metabolite found in all *Leucaena* and *Mimosa* species. We also describe the methods used for characterization of the recombinant mimosinase.

Key words Mimosine, Mimosinase, Leucaena, 3-Hydroxy-4-pyridone, C-N lyase, Aminotransferase, Secondary metabolite, Heterologous expression

1 Introduction

Plants are the best synthetic chemists; they can synthesize a vast array of secondary metabolites based on innumerable skeletal structures and functional group combinations [1, 2]. Although technological advances in genomics, transcriptomics, proteomics, and metabolomics have been instrumental in deciphering metabolic pathways for some of the plant secondary metabolites (PSMs), the metabolic pathways for the majority of PSMs are still a challenge for biochemists and molecular biologists. One such PSM for which the complete metabolic pathway is still unknown is mimosine, an N-heterocyclic nonprotein aromatic amino acid (Fig. 1) from the tree legume *Leucaena leucocephala* (leucaena).

Leucaena is an important agroforestry tree in the tropics and is considered as a promising fodder because of its protein-rich foliage [3–5]. However, its use as a fodder is limited as its foliage contains a large amount of mimosine, which is toxic to prokaryotes and

Arthur Germano Fett-Neto (ed.), *Biotechnology of Plant Secondary Metabolism: Methods and Protocols*, Methods in Molecular Biology, vol. 1405, DOI 10.1007/978-1-4939-3393-8_7, © Springer Science+Business Media New York 2016

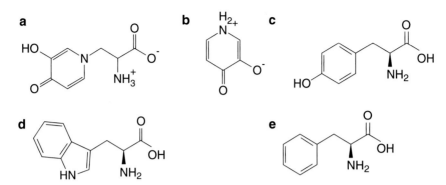

Fig. 1 Chemical structures of (**a**) mimosine, (**b**) 3-hydroxy-4-pyridone, (**c**) ʟ-tyrosine, (**d**) ʟ-tryptophan, and (**e**) ʟ-phenylalanine

eukaryotes [6, 7]. The biochemical basis of mimosine toxicity lies in its ability to (1) chelate divalent metallic ions, and (2) form a stable complex with pyridoxal-5′-phosphate (PLP). As a consequence, mimosine can limit the amount of metallic ions and PLP for use as cofactors and coenzymes, and thereby inactivate a wide variety of enzymes, which are dependent on divalent metallic ions or PLP [8].

Genomic and transcriptomic approaches are often used for identifying the genes involved in the metabolism of PSMs. Once a candidate gene is identified, it is expressed in a heterologous system and the recombinant enzyme is functionally characterized in vitro [9]. We also identified a candidate gene for mimosine catabolism using interspecies suppression subtractive hybridization, a transcriptomic approach [10]. In this chapter, we describe the method of leucaenamimosinaseheterologous expression in *E. coli* and in vitro functional characterization of the recombinant enzyme.

2 Materials

1. Mature seeds of *Leucaena leucocephala* Lam (de Wit) were collected from University of Hawaii research station, Waimanalo, Honolulu.

2. Half-strength Murashige and Skoog basal media: Add 2.2 g of Murashige and Skoog basal medium with Gamborg's vitamins (Sigma-Aldrich) and 1.5 % sucrose (w/v) in 950 mL Milli-Q water. Adjust pH to 5.8 and make up the volume to 1 L. Add 0.65 % phytagel (w/v) and autoclave on liquid cycle for 20 min at 15 psi.

3. LB media: Dissolve 10 g tryptone, 5 g yeast extract, and 10 g NaCl in 950 mL Milli-Q water. Adjust the pH of the medium to 7.0 and bring volume up to 1 L. Autoclave on liquid cycle for 20 min at 15 psi.

4. LB agar plates with antibiotics: Dissolve 10 g tryptone, 5 g yeast extract, and 10 g NaCl in 950 mL Milli-Q water. Adjust the pH of the medium to 7.0 and bring volume up to 1 L. Add 15 g/L agar and autoclave on liquid cycle for 20 min at 15 psi. After autoclaving, cool to ~55 °C, and add appropriate antibiotic. Gently swirl the media to mix the antibiotic and pour into petri dishes. Let solidify the media, then invert, and store at +4 °C.

5. LB agar plates for blue white screening: Add 100 μL of 0.1 M IPTG and 2 % X-gal to the surface of an LB/antibiotic plate. Spread IPTG and X-gal using a bacterial spreader and allow fluid to absorb for at least 30 min prior to bacterial plating.

3 Methods

3.1 Identification and Isolation of Mimosine Degradation Gene from Leucaena cDNA Library

1. Leucaena cDNA library sequences obtained from interspecies suppression subtractive hybridization [10] were screened for the presence of C-N lyase using the BLASTX tool from the NCBI BLAST homepage (*see* **Notes 1** and **2**).

2. The reference protein database was selected for BLAST search. The cDNA fragment that showed homology to C-N lyase was amplified to full-length cDNA by performing RNA ligase-mediated rapid amplification of cDNA ends (RLM-RACE) as described in Subheading 3.3.

3.2 RNA Extraction from Leucaena Seedlings

1. Place mature seeds of leucaena in a glass container and cover them with concentrated sulfuric acid for 5 min in glass bottle, followed by rinsing 8–10 times with cold water (*see* **Note 3**).

2. Surface sterilize the acid-scarified seeds by treating them with 3 % v/v sodium hypochlorite for 10 min. Rinse the seeds with sterile double-distilled water for 8–10 times to remove any traces of sodium hypochlorite (*see* **Note 4**).

3. Place surface sterilized seeds on petri dishes containing half-strength Murashige and Skoog media (MS media), seal the plates with parafilm, and incubate at 28 °C in the dark for 1 week.

4. Transfer germinated seedlings to magenta boxes containing MS media.

5. Collect young shoots of leucaena from 8-week-old seedlings and immediately freeze in liquid nitrogen (*see* **Note 5**).

6. The frozen tissues are finely pulverized in liquid nitrogen using mortar and pestle.

7. For RNA extraction, 1 mL of TRI Reagent is added to 100 mg of pulverized tissue in a 1.5 mL tube.

8. Vigorously vortex the tube to homogenize the tissue (*see* **Note 6**).

9. Transfer the clear supernatant to a fresh tube and allow the sample to stand for 5 min at room temperature (*see* **Note 7**).

10. For phase separation, add 0.2 mL of chloroform to the sample and close the tube tightly (*see* **Note 8**).

11. Vigorously vortex the tube for 30 s and allow to stand for 15 min at room temperature.

12. Centrifuge the resultant mixture at $12,000 \times g$ for 15 min at 4 °C. Centrifugation separates the mixture into three phases: a red organic phase (containing protein), an interphase (containing DNA), and a colorless upper aqueous phase (containing RNA).

13. Transfer the aqueous phase containing RNA to a fresh tube, add 0.5 mL of 2-propanol, and gently mix by inverting the tube 5–6 times.

14. Allow the resultant mixture to stand for 10 min at room temperature; next, centrifuge at $12,000 \times g$ for 10 min at 4 °C. The RNA precipitate will form a pellet on the side and bottom of the tube.

15. Discard the supernatant and add 1 mL of 75 % ethanol to the RNA pellet. Vortex the sample and then centrifuge at $7500 \times g$ for 5 min at 4 °C (*see* **Note 9**).

16. Air-dry the RNA pellet for 5–10 min (*see* **Note 10**).

17. Dissolve the pellet in 35 μL RNase-free water (*see* **Note 11**).

3.3 Rapid Amplification of cDNA Ends (RACE) of Mimosinase cDNA: 5′-RACE

1. In an RNase-free tube, add 10 μg leucaenaRNA and 2 μL each of calf intestine alkaline phosphatase (CIP) and 10× CIP buffer.

2. Make up the volume of the reaction to 20 μL using nuclease-free water.

3. Gently mix the reaction using pipette and incubate at 37 °C for 1 h (*see* **Note 12**).

4. Following 1 h of CIP treatment, add ammonium acetate solution (15 μL), nuclease-free water (115 μL), and acid phenol:chloroform to the reaction (*see* **Note 13**).

5. Vortex the tube vigorously for 30 s and then centrifuge for 5 min at $17,949 \times g$ at room temperature in a benchtop microfuge.

6. Transfer the resultant aqueous phase (top layer) to a new RNase-free tube and add 150 μL isopropanol. Vortex the mixture vigorously and keep on ice for 10 min.

7. Centrifuge the mixture at $17,949 \times g$ at 4 °C for 20 min in a benchtop microfuge.

8. For washing, add 0.5 mL ice-cold ethanol (70 %) to RNA pellet and briefly vortex followed by centrifugation at 17,949 × g at 4 °C for 5 min in a benchtop microfuge.

9. Discard ethanol and then air-dry the RNA pellet.

10. Resuspend RNA pellet in 11 μL nuclease-free water.

11. In an RNase-free microcentrifuge tube, add 5 μL CIP-treated RNA (from **step 10**), 1 μL 10× TAP buffer, 2 μL tobacco acid pyrophosphatase (TAP), and 2 μL nuclease-free water.

12. Gently mix the reaction mixture using pipette and incubate at 37 °C for 1 h (*see* **Note 14**).

13. In a new RNase-free microfuge tube, add 2 μL CIP/TAP-treated RNA (from **step 12**), 1 μL 5′- RACE adapter, 1 μL 10× RNA ligase buffer, 2 μL T4 RNA ligase (2.5 U/μL), and 4 μL nuclease-free water (*see* **Note 15**).

14. Gently mix the reaction mixture using pipette and incubate at 37 °C for 1 h (*see* **Note 14**).

15. In a new RNase-free microfuge tube, set up reverse transcription reaction by adding 2 μL ligated RNA (from **step 14**), 4 μL dNTP mix, 2 μL random decamers, 2 μL 10× RT buffer, 1 μL RNase inhibitor, 1 μL M-MLV reverse transcriptase, and 8 μL nuclease-free water.

16. Gently mix the reaction mixture using pipette and incubate at 42 °C for 1 h (*see* **Note 14**).

17. In a nuclease-free PCR tube, the outer PCR reaction is set up by adding 1 μL reverse transcription reaction (from **step 16**), 10 μL 5× Phusion HF buffer, 4 μL dNTP mix, 2 μL seq3-F primer (5′-RACE gene-specific primer; 10 μM), 2 μL 5′-RACE outer primer, 0.25 μL Phusion DNA polymerase (2 U/μL), and 30.75 μL nuclease-free water (*see* **Note 16**). Primer sequences are given in Table 1.

18. Gently mix the reaction mixture in the tube and briefly spin to bring the contents to the bottom.

19. Set up the PCR parameters as follows: initial denaturation at 98 °C for 30 s; 30 cycles of 98 °C for 10 s, 60 °C for 15 s, and 72 °C for 40 s; and a final extension of 5 min at 72 °C (*see* **Note 17**).

20. In a nuclease-free PCR tube, the inner PCR reaction is set up by adding 1 μL outer PCR products (from **step 19**), 10 μL 5× Phusion HF buffer, 4 μL dNTP mix, 2 μL seq3-F primer (5′-RACE gene-specific primer; 10 μM), 2 μL 5′-RACE inner primer, 0.25 μL Phusion DNA polymerase (2 U/μL), and 30.75 μL nuclease-free water.

21. Gently mix the reaction solution in the tube and briefly spin to bring contents to the bottom.

Table 1
Primers and adapters used in this study (reproduced from ref. 8 with permission from American Society of Plant Biologists)

Primers/adapters	Description or sequence (5′–3′)	Source/reference
seq3-5′-RACE	AGCACCATATGCATGGGCTATCCT (for 5′-RACE of seq3 cDNA fragment)	This study
seq3-3′-RACE	TTGCTGGAAACAGCAGTTGCATGG (for 3′-RACE of seq3 cDNA fragment)	This study
5′-RACE adapter	GCUGAUGGCGAUGAAUGAACACUGCGUUUGCUGGC UUUGAUGAAA	First choice RLM-RACE kit (Ambion)
5′-RACE outer primer	GCTGATGGCGATGAATGAACACTG	
5′-RACE inner primer	GCGGATCCGAACACTGCGTTTGCTGGCTTTGATG	
3′-RACE adapter	GCGAGCACAGAATTAATACGACTCACTATAGGT12VN	
3′-RACE outer primer	GCGAGCACAGAATTAATACGACT	
T7 promoter primer-f	TAATACGACTCACTATAGGG (used for sequencing insert in pET-14b vector)	This study
T7 terminator primer-r	GGGTTATGCTAGTTATTGCT (used for sequencing insert in pET-14b vector)	This study
M13-F	GTTTTCCCAGTCACGAC (used for sequencing insert in pGEM-T easy vector)	This study
M13-R	CAGGAAACAGCTATGAC (used for sequencing insert in pGEM-T easy vector)	This study

22. The PCR parameters are set as in the outer PCR reaction.

23. Load 5 μL of PCR products on 1.5 % agarose gel containing 1 μg/mL ethidium bromide and visualize on a UV transilluminator.

3.4 Rapid Amplification of cDNA Ends (RACE) of Mimosinase cDNA: 3′-RACE

1. A 20 μL reverse transcription reaction is set up in an RNase-free microfuge tube by adding 1 μg total RNA, 4 μL dNTP mix, 2 μL 3′-RACE adapter, 2 μL 10× RT buffer, 1 μL RNase inhibitor, and 1 μL M-MLV reverse transcriptase. The volume is adjusted to 20 μL by nuclease-free water.

2. Gently mix the reaction solution using pipette and incubate at 42 °C for 1 h (*see* **Note 14**).

3. In a nuclease-free PCR tube, the PCR reaction is set up by adding 1 μL reverse transcription reaction (from **step 2**), 10 μL 5× Phusion HF buffer, 4 μL dNTP mix, 2 μL seq3-R primer

(3′-RACE gene-specific primer; 10 μM), 2 μL 3′-RACE outer primer, 0.25 μL Phusion DNA polymerase (2 U/μL), and 30.75 μL nuclease-free water (*see* **Note 16**).

4. Gently mix the reaction solutions in the tube and briefly spin to bring the contents to the bottom.

5. Set the PCR parameters as follows: initial denaturation at 98 °C for 30 s; 30 cycles of 98 °C for 10 s, 60 °C for 15 s, and 72 °C for 40 s; and a final extension of 5 min at 72 °C (*see* **Note 17**).

6. Load 5 μL of PCR products on 1.5 % agarose gel containing 1 μg/mL ethidium bromide and visualize on a UV transilluminator.

3.5 Cloning of RACE Products

1. In two independent ligation reactions, 5′- and 3′- RACE products are ligated to the pGEM®-T Easy Vector by adding 1 μL of PCR product, 1 μL pGEM®-T Easy Vector (50 ng), 5 μL 2× rapid ligation buffer, 2 μL nuclease-free water, and 1 μL T4 DNA ligase (3 Weiss units/μL) in a nuclease-free PCR tube (*see* **Note 18**).

2. Gently mix the reaction by pipetting and incubate overnight at 4 °C.

3. Chemically competent JM109 *E. coli* cells are thawed on ice and mixed by gentle flicking of tubes.

4. Transfer 50 μL of competent cells to the sterile tubes containing ligation reactions.

5. Add 5 μL of each of the ligation products to different vials of competent cells and keep the tubes on ice for 10 min followed by heat shock for 45 s at 42 °C in a water bath.

6. Immediately place tubes on ice for 2 min and then add 450 μL LB medium at room temperature.

7. Incubate the tubes at 37 °C with shaking at 180 rpm for 1 h.

8. From each transformation culture, spread 30 and 60 μL on LB/ampicillin/IPTG/X-Gal plates.

9. Incubate plates overnight at 37 °C.

10. White colonies are screened by colony PCR using M13 forward and reverse primers and plasmids from positive transformants are extracted and sequenced.

11. The sequences from 5′-RACE and 3′-RACE are assembled to obtain full-length mimosinase cDNA.

3.6 Sequence Analyses for Prediction of Signal Peptide and Presence of Rare Codons

1. To verify if the isolated cDNA actually encodes for a mimosine-degrading enzyme or not, it is important to express it in *E. coli*, purify the recombinant protein, and test the enzymatic activity. However, two major constraints in heterologous expression of a transgene are (1) presence of hydrophobic region, if any, and (2) presence of rare codons in the coding region of a trans-

gene. Hydrophobic regions, if present, usually make protein purification difficult by forming inclusion bodies [11, 12]; additionally, signal peptide is not essential for in vitro protein activity. Therefore, for efficient expression of the putative mimosinase gene in *E. coli*, it is possible to eliminate any potential signal peptide sequence and optimize the sequence based on the *E. coli* codon preferences.

2. The deduced amino acid sequence from the open reading frame (ORF) of full-length mimosinase cDNA is analyzed for the presence of any N-terminal signal peptide including chloroplast transit peptide (cTP), mitochondrial targeting peptide (mTP), or secretory pathway signal peptide (SP) using TargetP 1.1 server [13].

3. The TargetP 1.1 server predicted a 43-amino acid-long chloroplast transit peptide with a reliability class (RC) value of 2 at the N-terminus of the 443-amino acid sequence (*see* **Note 19**) (Table 2).

4. The 126 bp sequence for the predicted chloroplast transit peptide from 5′-end of the ORF, excluding the start codon, is eliminated for further analysis.

5. The remaining 1206 bp sequence is analyzed for the presence of *E. coli* rare codons based on its codon usage bias using GenScript Rare Codon Analysis Tool and a codon-optimized synthetic derivative of the ORF is obtained by replacing the rare codons with commonly used codons of *E. coli* (Fig. 2) (*see* **Note 20**). The sequence is also analyzed for the presence of potentially deleterious motifs that could have negative effect on protein expression in *E. coli* (*see* **Note 21**). The *E. coli* ribosome-binding site (AAGGAG), a potentially deleterious motif, is replaced with "AAAGAA" during codon optimization (Fig. 2).

6. In the 1206 bp synthetic ORF, a total of 258 out of 402 codons are changed by replacing 301 nucleotides to form synthetic mimosinase (*syn*-mimosinase). The comparison of the

Table 2
Prediction of subcellular localization of the putative mimosinase-encoding gene of *Leucaena leucocephala* using TargetP 1.1 server (reproduced from ref. 8 with permission from American Society of Plant Biologists)

| Protein | Length (amino acid) | Signal peptide neural network score on which final prediction is based | | | | Location | Reliability class[a] | Signal peptide length |
		Chloroplast transit peptide	Mitochondrial transit peptide	Secretory protein	Other			
Mimosinase	444	0.874	0.027	0.229	0.093	Chloroplast	2	43

[a]Reliability class (RC), from 1 to 5, where 1 indicates the strongest prediction

Fig. 2 Codon comparison of synthetic mimosinase (*Syn*-mimosinase) with wild-type mimosinase (Wt-mimosinase). *Bold* and *underlined texts* represent the replaced codons and the *shaded texts* represent the replaced nucleotide in the synthetic derivative of the wild-type mimosinase ORF. The *E. coli* ribosomal binding site (AAGGAG), which is a potentially deleterious site for heterologous expression of a transgene in *E. coli*, is underlined and highlighted in *yellow color*

coding region of wild-type mimosinase and synthetic mimosinase is shown (Fig. 2).

7. *Bam*HI restriction site is introduced at both the ends of the synthetic gene to facilitate its cloning in the expression vector.

3.7 Preparation of Syn-Mimosinase Expression Construct

1. Kits and reagents mentioned in this section are used according to the manufacturer's instruction unless otherwise stated.

2. Digest the codon-optimized *syn*-mimosinase and the expression vector pET-14b with *Bam*HI.

3. Purify the *Bam*HI-digested insert using QIAquick PCR Purification Kit.

4. Dephosphorylate the 5′-phosphates from the *Bam*HI-digested vector using shrimp alkaline phosphatase.

5. Load the digested and dephosphorylated vector on 1 % agarose gel containing 1 μg/mL ethidium bromide and visualize on a UV transilluminator.

6. Excise the band for digested pET14-b using clean blade and gel purify using QIAquick gel extraction kit.

7. Ligate the purified insert (from **step 2**) and vector (from **step 5**) using T4 ligase (*see* **Note 22**).

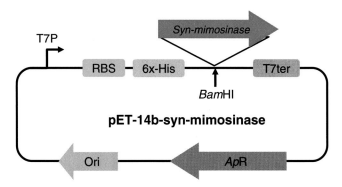

Fig. 3 Cloning of *syn-mimosinase* ORF in *E. coli*. The expression construct for *syn-mimosinase* ORF expression was assembled by cloning the codon-optimized synthetic mimosinase ORF lacking coding region for chloroplast signal peptide, at the *Bam*HI restriction site of pET-14b vector in the sense orientation

8. Clone the ligation reaction in chemically competent JM109 *E. coli* cells as previously described in Subheading 3.3, **step 3**.

9. Verify the sequence of the expression construct for correct orientation and frame (Fig. 3).

3.8 Heterologous Expression of Syn-Mimosinase in E. coli and Purification of the Encoded Enzyme

1. Transform the expression construct containing *syn*-mimosinase in correct orientation and frame into BL21 (DE3) pLysS *E. coli* cells (*see* **Note 23**). Plate the transformation reaction on LB agar plates containing chloramphenicol and ampicillin at the final concentration of 50 μg/mL and 100 μg/mL, respectively.

2. Inoculate the transformants in 2 mL LB medium containing chloramphenicol and ampicillin at the final concentration of 50 μg/mL and 100 μg/mL, respectively (*see* **Note 24**).

3. Incubate the culture at 37 °C overnight in an incubator shaker with a shaking speed of 250 rpm.

4. From the overnight culture, add an 100 μL aliquot to 2 mL of fresh LB medium containing no antibiotic. Incubate at 37 °C with shaking at 250 rpm for 2 h.

5. After 2-h incubation, pipette 200 μL of the culture into a clean microfuge tube and keep on ice. This sample is labeled as 0-h induced (non-induced) sample.

6. To the remaining culture add isopropyl β-D-1-thiogalactopyranoside (IPTG) to a final concentration of 1 mM for inducing the expression of *syn*-mimosinase.

7. Incubate the culture with shaking at 220–250 rpm at 37 °C for 10 h, collect 200 μL samples from the induced culture at 1, 2, 4, 6, 8, and 10 h, and keep them on ice.

8. Centrifuge samples at $1702 \times g$ for 10 min at 4 °C and resuspend the pellet in 1× SDS-PAGE loading buffer.

(a)

Induction time (Hours)

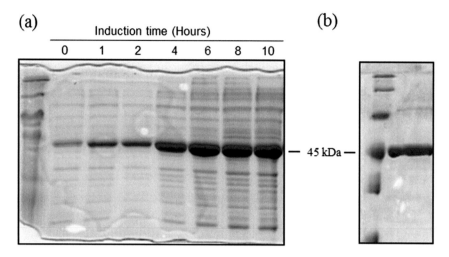

0 1 2 4 6 8 10

(b)

— 45 kDa —

Fig. 4 Heterologous expression and purification of recombinant mimosinase. (**a**) Total protein from BL21 (DE3) pLysS *E. coli* cells containing expression construct for mimosinase. The expression of recombinant protein was induced for different times. (**b**) The purified recombinant protein from 6-h induced BL21 (DE3) pLysS *E. coli* cells containing expression construct for mimosinase

9. For total protein extraction, samples in 1× SDS-PAGE loading buffer are heated at 95 °C for 5 min.

10. Analyze the total protein samples on SDS-PAGE for optimum induction time (Fig. 4a).

11. Purify the recombinant protein, using MagneHis Protein Purification System, from a separate culture induced for 6 h (Fig. 4b) (*see* **Note 25**).

3.9 Activity Assay for Recombinant Mimosinase

1. Carry out each reaction in three replications.

2. The catalytic activity of the purified recombinant mimosinase is tested by using mimosine as the substrate. Tris–HCl (0.1 M), pH 7.5 is used as the reaction buffer.

3. In 1 mL enzymatic reaction, add 0.016 mg of the purified enzyme and 1 mM mimosine, followed by incubation at 37 °C for 1 h. Heat-inactivated recombinant mimosinase is used in control reactions.

4. Stop reaction by boiling the tubes containing reaction mixture for 3 min.

5. Filter the reaction mixture through 0.2 μm filters and assay by HPLC.

6. For HPLC analysis, use a C18 column (4.6 × 250 mm; Dionex Acclaim 120), an isocratic solvent system of 0.02 M *o*-phosphoric acid with a flow rate of 1 mL/min, and UV detection by photodiode array (200–400 nm).

7. Different concentrations of commercially available mimosine and chemically synthesized 3-hydroxy-4-pyridone (synthetic 3H4P) can be used as standards. Synthetic 3H4P was originally obtained from Dr. Edward J. Behrman (Ohio State University), who previously described an improved method for the synthesis of 3H4P [14].

8. Record the peak area of standards and test samples (*see* **Note 26**).

9. Plot a standard curve from the peak area of known concentrations of synthetic 3H4P, which is then used to quantify the amount of product formed from the peak area of test samples (Fig. 5).

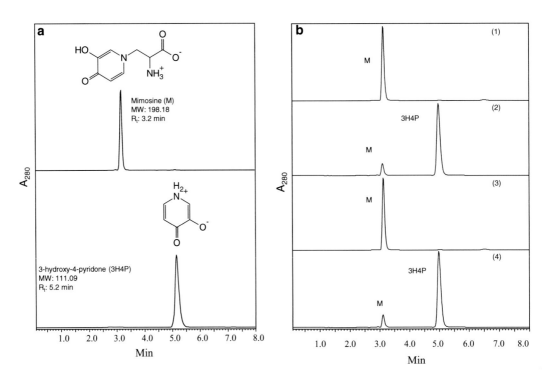

Fig. 5 HPLC chromatograms of standards and test samples in mimosine degradation assays. (**a**) The chromatograms represent mimosine and 3H4P that were used as standards. The chromatograms of mimosine and 3H4P had a retention time of 3.2 min and 5.2 min, respectively. (**b**) HPLC chromatograms of mimosine degradation assays: (1) reaction in which heat-inactivated recombinant enzyme was added and exhibited a single peak of unused substrate, mimosine; (2) reaction catalyzed by functionally active recombinant enzyme in the absence of exogenously added α-KG and PLP exhibited a large peak of mimosine degradation product with the same retention time as that of 3H4P; (3) reaction catalyzed by functionally active recombinant enzyme in the presence of 50 μM hydroxylamine had only one peak of unutilized substrate; (4) functionally active recombinant enzyme-catalyzed reaction in the presence of inhibitor, 50 μM hydroxylamine, restored the enzyme activity when supplemented with 0.1 μM PLP and showed large peak of mimosine degradation product with the same retention time as that of 3H4P (reproduced from [8] with permission from American Society of Plant Biologists)

3.10 Characterization of Recombinant Mimosinase: Test for Optimum pH and Optimum Temperature

1. The mimosine degradation assay for these tests is performed as described in Subheading 3.7 unless otherwise stated.

2. For identifying the optimum temperature for the catalytic activity of recombinant mimosinase, the mimosine degradation assay is performed at different temperatures (4, 22, 30, 37, 45, 55, and 65 °C) while keeping the pH constant at 7.5 (Fig. 6).

3. For determining optimum pH, the mimosine degradation assay is performed at different pH (pH 3, 5, 6, 7, 7.5, 8, 8.5, 10, and 12) and the reactions are incubated at 37 °C (Fig. 6).

4. For studying the thermal stability of recombinant mimosinase, the purified enzyme is preincubated in the reaction buffer at different temperatures, including 4, 37, 40, 50, 55, 60, 65, and 70 °C for 30 min. Following preincubation, the substrate is added and the reaction mixture is incubated at 37 °C for 1 h (Fig. 6).

5. All subsequent assays described in the next sections are carried out at optimum temperature (37 °C) and optimum pH (pH 8).

6. The product formed in each reaction is analyzed and quantified using HPLC as described in previous sections.

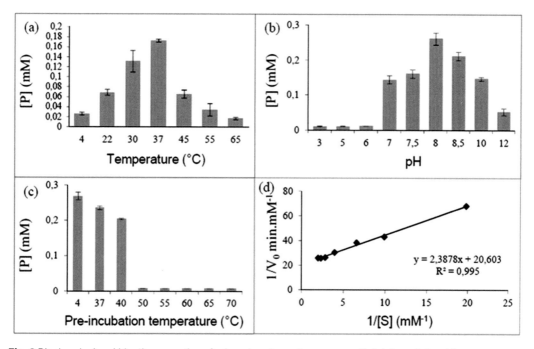

Fig. 6 Biochemical and kinetic properties of mimosine-degrading enzyme. Catalytic activity of the enzyme was determined as the product formed [P] at (**a**) different temperatures; (**b**) different pHs; and (**c**) different preincubation temperatures. (**d**) Estimation of kinetic parameters of the enzyme obtained by plotting Lineweaver-Burk plot of initial velocities at different substrate concentrations (reproduced from [8] with permission from American Society of Plant Biologists)

3.11 Characterization of Recombinant Mimosinase: Test for Aminotransferase and Lyase

1. To determine whether the recombinant enzyme is an aminotransferase, the mimosine degradation assay is performed in both the presence and absence of 1 mM α-KG and 20 mM PLP in the reaction buffer (*see* **Note 27**).

2. To test if recombinant mimosinase is a lyase, the PLP dependence of the recombinant enzyme by supplementing the reaction buffer with hydroxylamine, which is a potent inhibitor of PLP-dependent enzymes, is checked.

3. PLP in different concentrations (0.01–50 mM) is added to the reaction buffer containing enzyme but no substrate.

4. The reaction mixture is incubated for 5 min at room temperature before adding the substrate.

5. In a separate set of reactions, the reaction mixture containing 50 μM hydroxylamine and enzyme is supplemented with 0.1–20 μM PLP and again incubated for 5 min at room temperature followed by the addition of substrate (*see* **Note 28**).

6. After addition of the substrate the reaction is performed and analyzed as described in Subheading 3.7 (Fig. 5).

3.12 Characterization of Recombinant Mimosinase: Test for Competitive Inhibition Using Aromatic Amino Acids as Structural Analog of Mimosine

1. The recombinant mimosinase is also analyzed for any possible competitive inhibition by mimosine structural analogs, including the aromatic amino acids, L-Tyr, L-Trp, and L-Phe.

2. The mimosine degradation assay for these tests is performed as described in previous sections.

3. The inhibitor-to-substrate ratio is adjusted to 1:1, 2:1, and 3:1 in test reactions.

4. For control reaction, only substrate is used in the reaction.

5. The product formed in each reaction is analyzed and quantified using HPLC as described in previous sections (Fig. 7).

3.13 Characterization of Recombinant Mimosinase: Kinetic Properties of Recombinant Mimosinase

1. The enzymatic assays for determining the kinetic properties are performed under optimal conditions as described in previous sections unless otherwise stated.

2. For determining initial velocity of recombinant mimosinase, 0.05, 0.1, 0.15, 0.25, 0.35, and 0.5 mM mimosine is used as substrate.

3. The reactions are conducted for different time points including 0, 2, 3, 5, 10, 20, and 30 min.

4. The rate of reaction is linear between 0 and 2 min. Therefore, for determining kinetic properties, these two time points are used.

5. The amount of product formed at 0 and 2 min in reactions with different substrate concentrations was determined by HPLC.

6. The initial velocities for recombinant mimosinase with different substrate concentrations are calculated as the slopes formed by the amount of product formed at 0 and 2 min.

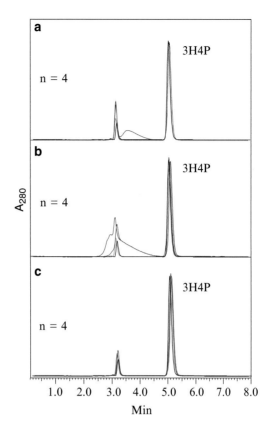

Fig. 7 HPLC chromatograms of mimosine degradation assays performed in the presence of structural analogs of mimosine, including (**a**) L-tyrosine; (**b**) L-phenylalanine; and (**c**) L-tryptophan. Each of these structural analogs was used in three different concentrations (1, 2, and 3 mM) in separate reactions with 1 mM mimosine. The reaction in the absence of structural analog served as the control. *n* = 4 represents four reactions; 1–3: structural analog at concentrations 1, 2, and 3 mM; 4: reaction without structural analog. Each of the reactions had 1 mM of the mimosine as the substrate. The product formed in control and test reactions showed no difference in the amount of product formed. This shows that none of the structural analogs competitively inhibits mimosinase-catalyzed reaction (reproduced from [8] with permission from American Society of Plant Biologists)

7. The calculated initial velocities and the substrate concentrations present in reactions are used for plotting a Lineweaver-Burk plot, and the V_{max} and K_m for recombinant mimosinase are calculated from linear regression of the plot (Fig. 6).

8. The apparent mol of enzyme active sites per mg of the enzyme ($[E]_T$) is calculated assuming that there is only one active site per enzyme molecule.

9. The turnover number (K_{cat}) for recombinant mimosinase is calculated using V_{max} and $[E]_T$ (*see* **Note 29**).

4 Notes

1. We expected the mimosine-degrading enzyme gene to encode a C-N lyase because a C-N lyase from leucaena has been previously isolated and shown to have mimosine-degrading activity [15]. Additionally, we have also shown that a mimosine degradation gene, namely *midD*, from leucaena symbiont *Rhizobium* sp. strain TAL1145 encodes a C-N lyase that degrades mimosine into 3H4P, pyruvate, and ammonia [16]. Therefore, we decided to screen for a C-N lyase from the leucaena cDNA library sequences.

2. When analyzing multiple sequences from a cDNA library, submitting multiple query sequences in a single BLAST search would be faster. However, one should make sure to submit sequences in FASTA format with sequence identifier.

3. Acid scarification using concentrated sulfuric acid helps in breaking dormancy of mature seeds with hard seed coat by modifying (thinning) the seed coat. However, sulfuric acid should be used with extreme caution.

4. The surface sterilization of seeds is performed under aseptic conditions in laminar flow hood.

5. Use only cryogenic container specifically designed for liquid nitrogen and do not put any heavy item over the lid of liquid nitrogen container as any interference with the venting of gas may result in explosion. Make sure to wear full-length apron and cover your feet with shoes. Always protect your eyes with safety glasses or a face shield while using liquid nitrogen. The gas released from the liquid is also extremely cold and any exposure to the cold gas can damage delicate tissues such as eyes. Protect hands all the time with gloves. Loose-fitting gloves are better as they could be thrown off if liquid nitrogen accidently goes inside them. Vaporization of liquid nitrogen rapidly reduces the oxygen percentage. The reduction of oxygen percentage below 19.5 % causes oxygen deprivation which may cause suffocation-related death as liquid nitrogen does not have good warning properties. Therefore, it is very important to use liquid nitrogen in well-vented area. Also do not store liquid nitrogen containers in cold room. Cold rooms have no air exchanges and elevated nitrogen in the air may cause sudden suffocation to someone who enters the room.

6. In TRI Reagent, samples can be stored at −70 °C for up to 1 month and the homogenate was centrifuged at $12,000 \times g$ for 10 min at 4 °C. Although this step is considered as optional in many product (TRI Reagent) manuals, this step is important for RNA quality of leucaena. Centrifugation of the homogenate helps removing the insoluble material, such as extracellular membranes, polysaccharides, and high-molecular-mass DNA.

7. This step ensures complete dissociation of nucleoprotein complexes.

8. The chloroform should not contain isoamyl alcohol or other additives. Alternatively, 1-bromo-3-chloropropane can be used, which is less toxic than chloroform, and its use for phase separation decreases the possibility of contaminating RNA with DNA.

9. If the RNA pellets float, centrifuge at $12,000 \times g$.

10. Do not let the RNA pellet dry completely, as this will greatly decrease its solubility. Do not dry the RNA pellet by centrifugation under vacuum (Speed-Vac).

11. To facilitate resuspension of RNA, the solution is incubated at 60 °C for 15 min.

12. CIP treatment removes $5'$-PO_4 from degraded mRNA, rRNA, tRNA, and DNA. The $5'$-cap present in intact mRNA remains unaffected by CIP treatment.

13. Acid phenol:chloroform gives better result compared to ordinary phenol:chloroform.

14. Reaction can be stored at −20 °C at this stage without compromising the result.

15. For better ligation efficiency, warm the 10× RNA ligase buffer quickly by rolling the tube between gloved hands and vortex at medium speed to resuspend any precipitate. The 10× RNA ligase buffer contains ATP; therefore, do not heat the buffer over 37 °C.

16. Assemble the PCR reaction on ice. A minus-template control reaction is also set up in which template is replaced with nuclease-free water. This reaction is important to test any possible contamination in the PCR reagents.

17. Preheat the PCR machine to 98 °C before putting the reaction tube in the machine.

18. The 2× rapid ligation buffer must be vortexed vigorously before use.

19. For signal peptide prediction "plant networks" may be selected in web interface of TargetP 1.1 server. The low RC value gives strong prediction of the transit peptide, indicating that the encoded protein may be localized in the chloroplast.

20. Transgenes are often poorly expressed in heterologous hosts because every organism has their own codon bias based on which several codons of a transgene may be rare codons for the heterologous host. The host has very limited supply of aminoacyl-tRNA for rare codons and, therefore, the presence of rare codons in a transgene may significantly influence heterologous expression yield by limiting the translation elongation rate [17, 18]. In some cases, the slowing down of ribosome

due to lack of aminoacyl-tRNA corresponding to rare codon results in premature termination of translation, a phenomenon called the polarity effect [19, 20].

21. For expressing a gene under the control of T7 promoter in *E. coli*, class I and II transcriptional termination sites and Shine-Dalgarno-like sequences in the open reading frame should be removed while optimizing the sequences. The termination sites may cause premature termination of transcription, whereas the presence of Shine-Dalgarno-like sequences may result in translation initiation at wrong place. The *E. coli* ribosomal binding site in the coding region may also compromise the protein expression.

22. The ligation of insert at *Bam*HI restriction site of pET14b positions *syn*-mimosinase sequence downstream of poly-His tag under the control of T7 promoter.

23. BL21 (DE3) pLysS *E. coli* cells are ideal for expressing a transgene under the control of a T7 promoter. BL21 cells are deficient in proteases Lon and OmpT. They also carry both the DE3 lysogen and the plasmid pLysS. The DE3 lysogen carries the T7 RNA polymerase gene and lacIq because of which the transformed plasmids containing T7 promoter-driven expression are repressed until IPTG induction of T7 RNA polymerase from a lac promoter. The pLysS plasmid encodes T7 phage lysozyme, an inhibitor of T7 polymerase which reduces and almost eliminates expression from transformed T7 promoter containing plasmids when not induced.

24. It is always better to test more than one colony as colony-to-colony variations in protein expression are possible.

25. The 6-h induction time is selected based on the optimum induction time from **step 10**.

26. Each standard and test samples are used in triplicates.

27. The BLASTP analysis of the deduced amino acid sequence of mimosinase ORF using reference protein database exhibits homology with proteins that belong to the aspartate aminotransferase superfamily. As not all the proteins in aminotransferase superfamily are in fact aminotransferases, and the closest homolog of mimosinase is a cystathionine β-lyase, it is likely that mimosinase is not an aminotransferase but a lyase. Therefore, to experimentally establish that mimosinase is not an aminotransferase, a catalytic reaction may be carried out in the presence and absence of α-KG, and PLP.

28. This set of assays should be performed to restore the enzyme activity if it is actually inhibited by hydroxylamine.

29. $K_{cat} = V_{max}/[E]_T$.

Acknowledgement

We thank James L Brewbaker (College of Tropical Agriculture and Human Resources, University of Hawaii at Manoa) and Edward J. Behrman for generously providing us leucaena seeds and synthetic 3H4P, respectively. This work was supported by the National Science Foundation Award No. CBET 08-27057 and partially by a HATCH grant (HAW00551-H). VSN was supported by IFP fellowship from the Ford Foundation for 3 years.

References

1. Wink M (2010) Introduction: biochemistry, physiology and ecological functions of secondary metabolites. In: Wink M (ed) Biochemistry of plant secondary metabolism, vol 40, Annual plant reviews. Wiley-Blackwell, Chichester, pp 1–19

2. Kroymann J (2011) Natural diversity and adaptation in plant secondary metabolism. Curr Opin Plant Biol 14:246–251

3. Garcia GW, Ferguson TU, Neckles FA, Archibald KAE (1996) The nutritive value and forage productivity of *Leucaena leucocephala*. Anim Feed Sci Technol 60:29–41

4. Soedarjo M, Borthakur D (1998) Mimosine, a toxin produced by the tree-legume *leucaena* provides a nodulation competition advantage to mimosine-degrading *rhizobium* strains. Soil Biol Biochem 30:1605–1613

5. Pal A, Negi VS, Borthakur D (2012) Efficient in vitro regeneration of *Leucaena leucocephala* using immature zygotic embryos as explants. Agroforest Syst 84:131–140

6. Lalande M (1990) A reversible arrest point in the late G1 phase of the mammalian cell cycle. Exp Cell Res 186:332–339

7. Soedarjo M, Hemscheidt TK, Borthakur D (1994) Mimosine, a toxin present in leguminous trees (*Leucaena* spp.), induces a mimosine-de grading enzyme activity in some *Rhizobium* strains. Appl Environ Microbiol 60:4268–4272

8. Negi VS, Bingham J-P, Li QX, Borthakur D (2014) A carbon-nitrogen lyase from *Leucaena leucocephala* catalyzes the first step of mimosine degradation. Plant Physiol 164:922–934

9. Facchini PJ, Bohlmann J, Covello PS, De Luca V, Mahadevan R, Page JE, Ro D-K, Sensen CW, Storms R, Martin VJ (2012) Synthetic biosystems for the production of high-value plant metabolites. Trends Biotechnol 30:127–131

10. Negi VS, Pal A, Singh R, Borthakur D (2011) Identification of species-specific genes from *Leucaena leucocephala* using interspecies suppression subtractive hybridisation. Ann Appl Biol 159:387–398

11. Fink AL (1998) Protein aggregation: folding aggregates, inclusion bodies and amyloid. Fold Des 3:R9–R23

12. Pal A, Negi VS, Khanal S, Borthakur D (2012) Immunodetection of curcin in seed meal of *Jatropha curcas* using polyclonal antibody developed against Curcin-L. Curr Nutr Food Sci 8:213–219

13. Emanuelsson O, Nielsen H, Brunak S, von Heijne G (2000) Predicting subcellular localization of proteins based on their N-terminal amino acid sequence. J Mol Biol 300:1005–1016

14. Behrman EJ (2009) Synthesis of 4-Pyridone-3-sulfate and an improved synthesis of 3-Hydroxy-4-Pyridone. Chem Cent J 3:1

15. Smith IK, Fowden L (1966) A study of mimosine toxicity in plants. J Exp Bot 17:750–761

16. Negi VS, Bingham J-P, Li QX, Borthakur D (2013) *midD*-encoded 'rhizomimosinase' from *Rhizobium* sp. strain TAL1145 is a C–N lyase that catabolizes L-mimosine into 3-hydroxy-4-pyridone, pyruvate and ammonia. Amino Acids 44:1537–1547

17. Curran JF, Yarus M (1989) Rates of aminoacyl-tRNA selection at 29 sense codons *in vivo*. J Mol Biol 209:65–77

18. Welch M, Villalobos A, Gustafsson C, Minshull J (2009) You're one in a googol: optimizing genes for protein expression. J R Soc Interface 6:S467–S476

19. Richardson JP (1991) Preventing the synthesis of unused transcripts by Rho factor. Cell 64:1047–1049

20. Proshkin S, Rahmouni AR, Mironov A, Nudler E (2010) Cooperation between translating ribosomes and RNA polymerase in transcription elongation. Science 328:504–508

Chapter 8

Production of Aromatic Plant Terpenoids in Recombinant Baker's Yeast

Anita Emmerstorfer-Augustin and Harald Pichler

Abstract

Plant terpenoids are high-value compounds broadly applied as food additives or fragrances in perfumes and cosmetics. Their biotechnological production in yeast offers an attractive alternative to extraction from plants. Here, we provide two optimized protocols for the production of the plant terpenoid *trans*-nootkatol with recombinant *S. cerevisiae* by either (I) converting externally added (+)-valencene with resting cells or (II) cultivating engineered self-sufficient production strains. By synthesis of the hydrophobic compounds in self-sufficient production cells, phase transfer issues can be avoided and the highly volatile products can be enriched in and easily purified from *n*-dodecane, which is added to the cell broth as second phase.

Key words Plant terpenoids, Yeast, *S. cerevisiae*, Cytochrome P450 enzymes, Biotransformation, (+)-Valencene synthase, t*HMG1*, Metabolic engineering

1 Introduction

Terpenes or terpenoids are primary constituents of essential oils of many types of plants and flowers and, hence, are of large industrial interest. Their isolation from natural resources is subjected to seasonal variations and often suffers from difficult removal of matrix-associated impurities. To date, some terpenoids are mainly produced via chemical synthesis that frequently involves toxic heavy metals, highly flammable compounds, or strong oxidants [1, 2]. Therefore, attention is being paid to the development of safe methods and innovative biotechnological solutions to reduce the ecological footprint [3–5]. The microbial synthesis of plant terpenoids has become an easy, comparably cheap, and viable alternative [6].

Terpenoids can be produced upon hydroxy-functionalization of cheap precursor terpenes by using highly specific membrane-anchored cytochrome P450 enzymes (CYPs) [7, 8]. CYP activity is tightly linked to a finely balanced system of NADPH cofactor recycling, oxygen supply, correct integration of heme iron into the active site of the CYP, and perfect interaction with the corresponding

Arthur Germano Fett-Neto (ed.), *Biotechnology of Plant Secondary Metabolism: Methods and Protocols*, Methods in Molecular Biology, vol. 1405, DOI 10.1007/978-1-4939-3393-8_8, © Springer Science+Business Media New York 2016

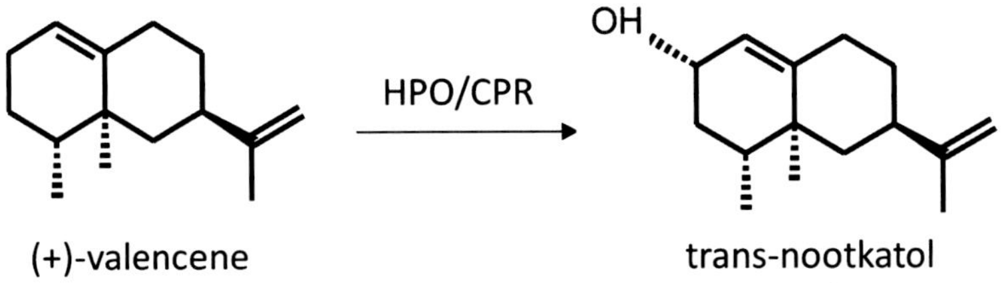

Fig. 1 Conversion of (+)-valencene to *trans*-nootkatol catalyzed by HPO and CPR

cytochrome P450 reductase (CPR) that functions as electron donor [9, 10]. Based on the system's complexity, the use of whole cellular systems is favored over cell lysates or purified enzymes to support cofactor supply and enable close interaction of the CYP enzyme with its corresponding reductase [11].

We have established and optimized protocols for the production of the model terpenoid *trans*-nootkatol in baker's yeast from (+)-valencene. We used a recombinant *S. cerevisiae* strain co-expressing CYP *Hyoscyamus muticus* premnaspirodiene oxygenase (HPO) and the cytochrome P450 reductase from *Arabidopsis thaliana* (CPR) to hydroxylate the C2 atom of (+)-valencene for the production of the high-value terpenoid compound *trans*-nootkatol [12–14] (Fig. 1). Phase transfer issues represent a considerable problem of whole-cell biotransformations of hydrophobic substrates. Thus we constructed an efficient *S. cerevisiae* strain capable of producing relevant amounts of (+)-valencene intracellularly [13].

In brief, yeast harbors the mevalonate pathway, which is the essential metabolic pathway necessary for the production of aroma components in plants. The yeast mevalonate pathway can be engineered to produce terpenes by implementing the corresponding terpene synthase, e.g., (+)-valencene synthase from Alaska cedar heartwood (*Callitropsis nootkatensis*) [15]. ValS [(+)-valencene synthase] forms (+)-valencene from farnesyl pyrophosphate (FPP). In order to increase the FPP pool in *S. cerevisiae*, a truncated version of *HMG1* was over-expressed. In numerous related projects, over-expression of t*HMG1* has successfully enhanced production of terpenoids [16–19]. Self-sufficient production of *trans*-nootkatol was performed in biphasic systems using *n*-dodecane for trapping the synthesized terpenoids [13]. This elegant method prevents toxicity or inhibitory effects of high concentrations of terpenoids on yeasts providing an online extraction step [16, 18, 20], because high concentrations have recently been described to hamper (+)-valencene biohydroxylation in *S. cerevisiae* [21]. Several other groups have published *S. cerevisiae* strains that self-sufficiently produce (+)-valencene and *trans*-nootkatol [15, 22]. Inspired by the cultivation protocol for *P. pastoris* [14], we optimized cultivation conditions by introducing a prolonged growth phase followed by milder induction conditions. This seemed to strengthen cells, allowing them to reach higher initial densities, followed by redirection of the metabolic flux to terpenoid products after induction.

2 Materials

2.1 Conversion of Externally Added Terpenes with S. cerevisiae Resting Cells

1. *S. cerevisiae* strain: W303 co-expressing CYP and CPR, e.g., HPO and CPR from a (+)-galactose inducible pYES2 co-expression vector harboring a uracil selection marker [13].

2. Amino acid drop-out powder (DOP): Grind equal amounts of adenine, lysine, tyrosine, histidine, leucine, and tryptophan with mortar and pestle to generate a homogeneous powder (*see* **Note 1**).

3. Synthetic defined-uracil plates (SD-ura plates): Dissolve 6.7 g of Difco™ yeast nitrogen base w/o amino acids (YNB) and 1 g of DOP in 900 mL of ddH$_2$O (*see* **Note 2**) and add 20 g of agar. Dissolve 20 g of glucose in 100 mL of ddH$_2$O. Autoclave both solutions and combine as soon as they are cooled to ~45 °C to pour plates.

4. Synthetic defined-uracil 2 % growth medium (SDG 2 % medium): Dissolve 6.7 g of Difco™ yeast nitrogen base w/o amino acids (YNB) and 1 g of DOP in 900 mL of ddH$_2$O (*see* **Note 2**). 20 g of glucose is dissolved in 100 mL ddH$_2$O. Autoclave and combine the solutions when they are cooled to room temperature.

5. Synthetic defined-uracil induction medium (SDI medium): Dissolve 6.7 g of Difco™ yeast nitrogen base w/o amino acids (YNB) and 1 g of DOP in 900 mL of ddH$_2$O (*see* **Note 2**). Dissolve 20 g of galactose and 7 g of raffinose (*see* **Note 3**) in 100 mL ddH$_2$O, sterile filter, and combine solutions.

6. Sterile 300 mL baffled flasks covered with aluminum foil.

7. Eppendorf BioPhotometer®.

8. 50 mL Falcon® tubes.

9. 50 mM potassium phosphate buffer, pH 7.4.

10. PYREX® tubes.

11. 100 mM (+)-valencene stock solution (prepare fresh before use): Weigh one drop of (+)-valencene (M = 204.35 g/mol) in a PYREX® tube and add DMSO to reach a final concentration of 100 mM (e.g., add 1 mL of DMSO to 2.04 mg of (+)-valencene). Add 1 % Triton®X-100 (*see* **Note 4**).

12. Ethyl acetate.

13. VXR basic Vibrax®.

14. Eppendorf 5810R centrifuge.

15. GC vials, inserts, crimp caps, and crimper.

16. GC-FID: Hewlett-Packard 6890 GC equipped with a flame ionization detector (FID), HP-5 column (cross-linked 5 % Ph-Me Siloxane; 10 m × 0.10 mm × 0.10 μm).

2.2 In Vivo Terpenoid Production with a Self-Sufficient S. cerevisiae Strain

1. Yeast strain: W303 expressing terpene synthase (+/− mevalonate pathway engineering), e.g., harboring integrated copies of ValS and t*HMG1*, and driving HPO/CPR expression from a co-expression vector [13].

2. Amino acid drop-out powder (DOP): Grind equal amounts of adenine, lysine, tyrosine, histidine, leucine, and tryptophane with mortar and pestle to generate a homogeneous powder.

3. Synthetic defined-uracil plates (SD-ura plates): Dissolve 6.7 g of Difco™ yeast nitrogen base w/o amino acids (YNB) and 1 g of DOP in 900 mL of ddH$_2$O (*see* **Note 2**) and add 20 g of agar. Dissolve 20 g of glucose in 100 mL of ddH$_2$O. Autoclave both solutions and combine as soon as they are cooled to ~45 °C to pour plates.

4. Synthetic defined-uracil 1 % growth medium (SDG 1 % medium): Dissolve 6.7 g of Difco™ yeast nitrogen base w/o amino acids (YNB) and 1 g of DOP in 900 mL of ddH$_2$O (*see* **Note 2**). Dissolve 10 g of glucose in 100 mL ddH$_2$O for a final glucose concentration of 1 %. Autoclave and combine the solutions as soon as they are cooled to room temperature.

5. Sterile 100 mL flasks covered with aluminum foil.

6. Eppendorf BioPhotometer®.

7. *n*-Dodecane.

8. Induction solution: 20 mg Galactose and 7 mg raffinose dissolved per mL of ddH$_2$O (*see* **Note 5**).

9. Supplementary solution: 0.167 g Yeast nitrogen base and 25 mg DOP dissolved per mL of sterile ddH$_2$O (*see* **Note 5**).

10. 50 mL Falcon® tubes.

11. Eppendorf 5810 R centrifuge.

12. GC vials, inserts, crimp caps, and crimper.

13. GC-FID: Hewlett-Packard 6890 GC equipped with a flame ionization detector (FID), HP-5 column (cross-linked 5 % Ph-Me Siloxane; 10 m × 0.10 mm × 0.10 μm).

3 Methods

3.1 Conversion of Externally Added Terpenes with Resting Cells (Fig. 2)

1. Freshly grow the *S. cerevisiae* strain harboring the cytochrome P450 and reductase co-expression plasmid on an SD-ura plate for 2–3 days [13].

2. Take a cell pellet about the size of a pinhead and inoculate 5 mL of SDG medium containing 2 % glucose in a 50 mL Falcon® tube. Shake cultures overnight at 170 rpm and 30 °C for 15–20 h.

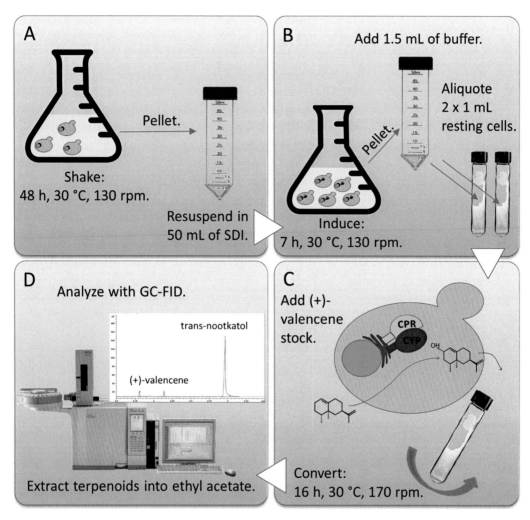

Fig. 2 Conversion of externally added terpenes with *S. cerevisiae* resting cells. After cultivating the yeast strain for 24 h in SDG 2 % medium, centrifuge the culture and resuspend the cell pellet in SDI medium (**A**). Induce co-expression of HPO and CPR by shaking cells for 7 h at 170 rpm and 30 °C. To generate resting cells, pellet 600 OD_{600} units of yeast cells and resuspend the pellet in 1.5 mL of 50 mM potassium phosphate buffer, pH 7.4, for two aliquots of 1 mL to be transferred to PYREX® tubes (**B**). Add 20 µL of 100 mM (+)-valencene stock solution each and let proceed bioconversions for 16 h (**C**). Extract terpenes and terpenoids with 500 µL of ethyl acetate on a VXR basic Vibrax® for 30 min. Separate phases by centrifugation and analyze organic phases by GC-FID (**D**)

3. Use overnight culture to inoculate 50 mL of SDG media in 300 mL baffled flasks to an OD_{600} of 0.1 and cultivate cells for 48 h at 130 rpm and 30 °C (*see* **Note 6**).

4. Transfer the cell suspension to a 50 mL Falcon® tube.

5. Spin down the cell suspension for 5 min at $1062 \times g$ in an Eppendorf 5810 R centrifuge and carefully resuspend the cell pellet in 50 mL of SDI medium (*see* **Note 7**).

6. Transfer the cell suspension to sterile 300 mL baffled flasks and induce cells for 7 h at 130 rpm and 30 °C (*see* **Note 8**).

7. Determine OD_{600} of induced cells with an Eppendorf BioPhotometer® and calculate the volume required to obtain 600 OD_{600} units. For example, if the end-OD_{600} of induced cells is 15, divide 600 by 15 to get the mL of cell volume needed. In this case, 40 mL of induced cells would be used to prepare resting cells.

8. For preparing resting cells, transfer the volume of induced cells equaling 600 OD_{600} units to a 50 mL Falcon® tube.

9. Centrifuge for 5 min at $1062 \times g$ in an Eppendorf 5810 R centrifuge and carefully resuspend the cell pellet in 1.5 mL of 50 mM potassium phosphate buffer, pH 7.4 (*see* **Note 9**).

10. Split the cell suspension into two equal aliquots in PYREX® tubes and add 20 μL (*see* **Notes 10** and **11**) of 100 mM (+)-valencene stock solution (*see* **Note 12**) to each aliquot.

11. Let biotransformations proceed for 16 h at 170 rpm and 30 °C (*see* **Note 13**).

12. Extract terpenoids into 500 μL of ethyl acetate with a VXR basic Vibrax® at room temperature and maximum speed for 30 min.

13. Separate phases by centrifugation at $2720 \times g$ for 15 min.

14. Transfer 70 μL of the upper ethyl acetate phase to GC-crimp vials (*see* **Note 14**).

15. GC-FID measurement: 1 μL of samples are injected in split mode (split ratio 30:1) at 250 °C injector temperature and 320 °C detector temperature with H_2 as carrier gas and a flow rate set to 0.4 mL/min in constant flow mode (49 cm/s linear velocity). Oven temperature program: 100 °C for 1 min, 20 °C/min ramp to 250 °C, and 45 °C/min ramp to 280 °C (0.5 min) [13].

16. For quantification, measure standard solutions with known concentrations of terpenoids with GC-FID and plot a calibration curve from the obtained results. Scale the abscissa in ng/μL of the terpenoid standard solutions. The ordinate represents the signal intensity measured.

3.2 Production of Terpenes and Terpenoids in a Self-Sufficient S. cerevisiae Strain

1. Freshly grow the *S. cerevisiae* strain harboring recombinant terpene synthase (+/– mevalonate pathway engineering), e.g., ValS, t*HMG1*, and CYP/CPR-co-expression plasmid [13], on an SD-ura plate for 2–3 days.

2. Inoculate 5 mL of SDG 1 % medium in a 50 mL Falcon® tube with a pin-head sized colony of the yeast strain. Shake overnight for 20 h at 25 °C and 170 rpm.

3. Determine OD_{600} and inoculate 50 mL of SDG 1 % medium (*see* **Note 15**) in 100 mL flasks to a starting OD_{600} of 0.1. Shake at 25 °C and 150 rpm for 24 h.

4. Add 1 mL of supplementary solution and 1 mL of induction solution (*see* **Note 16**) to the cell broth and directly overlay with 10 vol.-% (e.g., 5 mL) *n*-dodecane (Fig. 3). Cultivate at 25 °C and 150 rpm for another 24 h.

5. Add 1 mL of supplementary solution and 1 mL of induction solution and cultivate at 25 °C and 150 rpm for another 24 h.

6. Transfer the cell suspension including the organic phase to a 50 mL Falcon® tube.

7. Separate phases by centrifugation for 15 min at $2720 \times g$ Eppendorf 5810 R centrifuge.

8. Transfer 70 µL of the upper *n*-dodecane phase to a GC-crimp vial with insert (*see* **Note 17**).

9. GC-FID measurement: 1 µL of samples are injected in split mode (split ratio 30:1) at 250 °C injector temperature and 320 °C detector temperature with H_2 as carrier gas and a flow rate set to 0.4 mL/min in constant flow mode (49 cm/s linear velocity). Oven temperature program: 100 °C for 1 min, 20 °C/min ramp to 250 °C, and 45 °C/min ramp to 280 °C (0.5 min) [13] (*see* **Note 18**).

10. For quantification, measure standard solutions with known concentrations of terpenoids with GC-FID and plot a calibra-

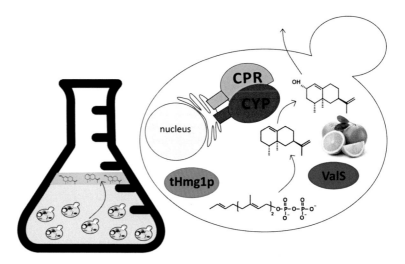

Fig. 3 In vivo terpenoid production with a self-sufficient *S. cerevisiae* strain. Upon over-expression of t*HMG1*, an increased intracellular pool of FPP is generated, which can be further converted to (+)-valencene with a highly specific valencene synthase (ValS). Subsequently, HPO/CPR convert the terpene to *trans*-nootkatol, which is enriched in *n*-dodecane overlaying the cell broth

tion curve from the obtained results. Scale the abscissa in ng/ µL of the terpenoid standard solutions. The ordinate represents the signal intensity.

4   Notes

1. Depending on the size of mortar and pestle take 2–4 g of each amino acid for preparing the DOP.

2. Stir the solution containing DOP and YNB prior to autoclaving for at least 2 h to achieve complete solubility of poorly soluble components.

3. The addition of 0.7 % raffinose to the SDI medium is optional, but it has shown to reduce induction stress by providing a growth advantage without affecting the induction process [23, 24].

4. Triton®X-100 is highly viscous and hard to pipette. Carefully take up the detergent by slowly handling the pipette. Flush the pipette tip multiple times with 100 mM (+)-valencene stock solution to quantitatively transfer all of the Triton®X-100.

5. Depending on the total number of yeast cultures to induce, prepare 50–100 mL of induction and supplementary solution. The solutions can be stored at 4 °C for up to 3 months.

6. Prolongation of the growth phase before induction was shown to be essential for high-level expression of CYP/CPR combinations [13] and is the most innovative aspect of this protocol.

7. Preheat SDI medium to 30 °C in an incubator or water bath to avoid any cellular cold shock response.

8. It is advisable to analyze activities of cells harvested after different time points of induction to find out the ideal time point providing the most active recombinant protein.

9. Depending on the pellet size of induced cells, the volume of 50 mM potassium phosphate buffer needs to be adjusted. Add 1.5 mL of 50 mM potassium phosphate buffer to 600 OD_{600} units to reach a final volume of 2 mL of resting cells.

10. Due to the high volatility of terpenes it is recommended to use stock solutions directly after preparation to avoid significant variations of substrate amounts applied for conversions.

11. DMSO and 1 % Triton®X-100 act as solubilizers for poorly water-soluble substrates. However, terpenes may separate from DMSO and form a second phase by time. To achieve the application of equal amounts of substrate per PYREX® tube, it is advisable to vortex the substrate stock solution every time before transferring the aliquots to resting cells for conversion.

12. Prior to conversion with *S. cerevisiae* resting cells, a toxicity test is recommended, because especially mono- and sesquiterpenoids are toxic for yeast cells. For toxicity studies, simply incubate your recombinant strain with aliquots of stock solutions containing different substrate concentrations. Compare the growth of treated versus untreated cells to estimate cellular toxicity [25, 26].

13. For activity assays with resting cells, seal PYREX® tubes loosely with screw caps and fix the caps with seal tape to maintain a balance between oxygen supply and substrate volatilization during conversion. Additionally, fix the PYREX® tubes in a rather steep angle to guarantee maximal turbidity. Make sure to maintain the same angle for all conversions.

14. Sometimes, the centrifugation step needs to be prolonged due to formation of a massive protein interlayer between the aqueous and organic phase. If phases cannot be separated properly, the addition of and extraction with another 500 µL of ethyl acetate are recommended to increase the volume of the upper organic phase. The additional volume must be considered in calculating terpenoid concentrations.

15. The amount of glucose needs to be reduced from 2 to 1 % to ensure depletion of the carbon source within 24 h of growth. Otherwise galactose induction will suffer from repression by residual glucose.

16. We initially applied 2.5 times higher galactose concentrations for induction, but this yielded only 50 % of the terpenoid productivity compared to the method described here. Due to the fact that galactose has a relatively high market price, lowering the galactose concentration has an additional economic benefit.

17. For disposal of two-phase cultures, gather the rest of the cultivations in a 2 L flask and let phases separate overnight. Remove the upper organic phase quantitatively and discard appropriately. Ensure solvent-free autoclaving of the aqueous cell broth.

18. The retention time of the analytes enriched in *n*-dodecane will change compared to the GC-FID spectra of samples extracted with ethyl acetate as described in Subheading 3.1. *n-Dodecane* causes all the compounds to elute from the column belatedly.

Acknowledgements

We thank Prof. Erich Leitner for providing help with GC-FID measurements. We also thank Martin Schürmann, Iwona Kaluzna, Monika Müller, Andreas Kolb, and Daniel Mink (DSM Chemical Technology R and D) for fruitful discussion. This work has been supported by the Federal Ministry of Science, Research and Economy (BMWFW), the Federal Ministry of Traffic, Innovation and Technology (bmvit), the Styrian Business Promotion Agency SFG, and the Standortagentur Tiroland ZIT—Technology Agency of the City of Vienna through the COMET—Funding Program managed by the Austrian Research Promotion Agency FFG.

References

1. Salvador JAR, Clark JH (2002) The allylic oxidation of unsaturated steroids by tert-butyl hydroperoxide using surface functionalised silica supported metal catalysts. Green Chem 4:352–356

2. Majetich G, Behnke M, Hull K (1985) A stereoselective synthesis of (+/-)-nootkatone and (+/-)-valencene via an intramolecular Sakurai reaction. J Org Chem 50:3615–3618

3. Fraatz MA, Berger RG, Zorn H (2009) Nootkatone—a biotechnological challenge. Appl Microbiol Biotechnol 83:35–41

4. Chang MCY, Keasling JD (2006) Production of isoprenoid pharmaceuticals by engineered microbes. Nat Chem Biol 2:674–681

5. Kirby J, Keasling JD (2009) Biosynthesis of plant isoprenoids: perspectives for microbial engineering. Annu Rev Plant Biol 60:335–355

6. Emmerstorfer A, Wriessnegger T, Hirz M, Pichler H (2014) Overexpression of membrane proteins from higher eukaryotes in yeasts. Appl Microbiol Biotechnol 98:7671–7698

7. Zhao Y-J, Cheng Q-Q, Su P, Chen X, Wang X-J, Gao W, Huang L-Q (2014) Research progress relating to the role of cytochrome P450 in the biosynthesis of terpenoids in medicinal plants. Appl Microbiol Biotechnol 98:2371–2383

8. Kitaoka N, Lu X, Yang B, Peters RJ (2015) The application of synthetic biology to elucidation of plant mono-, sesqui-, and diterpenoid metabolism. Mol Plant 8:6–16

9. Laursen T, Jensen K, Møller BL (2011) Conformational changes of the NADPH-dependent cytochrome P450 reductase in the course of electron transfer to cytochromes P450. Biochim Biophys Acta 1814:132–138

10. Jensen K, Møller BL (2010) Plant NADPH-cytochrome P450 oxidoreductases. Phytochemistry 71:132–141

11. Bernhardt R, Urlacher VB (2014) Cytochromes P450 as promising catalysts for biotechnological application: chances and limitations. Appl Microbiol Biotechnol 98:6185–6203

12. Takahashi S, Yeo Y-S, Zhao Y, O'Maille PE, Greenhagen BT, Noel JP, Coates RM, Chappel J (2007) Functional characterization of premnaspirodiene oxygenase, a cytochrome P450 catalyzing regio- and stereo-specific hydroxylations of diverse sesquiterpene substrates. J Biol Chem 282:31744–31754

13. Emmerstorfer A, Wimmer-Teubenbacher M, Wriessnegger T, Leitner E, Muller M, Kaluzna I, Schurmann M, Mink D, Zellnig G, Schwab H, Pichler H (2015) Overexpression of ICE2 stabilizes cytochrome P450 reductase in Saccharomyces cerevisiae and Pichia pastoris. Biotechnol J 10:623–635

14. Wriessnegger T, Augustin P, Engleder M, Leitner E, Muller M, Schurmann M, Mink D, Zellnig G, Schwab H, Pichler H (2014) Production of the sesquiterpenoid (+)-nootkatone by metabolic engineering of Pichia pastoris. Metab Eng 24:18–29

15. Beekwilder J, van Houwelingen A, Cankar K, van Dijk ADJ, de Jong RN, Stoopen G, Bouwmeester H, Achkar J, Sonke T, Bosch D (2014) Valencene synthase from the heartwood of Nootka cypress (Callitropsis nootkatensis) for biotechnological production of valencene. Plant Biotechnol J 12:174–182

16. Asadollahi MA, Maury J, Møller K, Nielsen KF, Schalk M, Clark A, Nielsen J (2008) Production of plant sesquiterpenes in Saccharomyces cerevisiae: effect of ERG9 repression on sesquiterpene biosynthesis. Biotechnol Bioeng 99:666–677

17. Farhi M, Marhevka E, Masci T, Marcos E, Eyal Y, Ovadis M, Abeliovioch H, Vainstein A (2011) Harnessing yeast subcellular compartments for the production of plant terpenoids. Metab Eng 13:474–481

18. Scalcinati G, Knuf C, Partow S, Chen Y, Maury J, Schalk M, Daviet L, Nielsen J, Siewers V (2012) Dynamic control of gene expression in Saccharomyces cerevisiae engineered for the production of plant sesquiterpene α-santalene in a fed-batch mode. Metab Eng 14:91–103

19. Ro D-K, Paradise EM, Ouellet M, Fisher KJ, Newman KL, Ndungu JM, Ho KA, Eachus RA, Ham TS, Kirby J, Chang MCY, Withers ST, Shiba Y, Sarpong R, Keasling JD (2006)

Production of the antimalarial drug precursor artemisinic acid in engineered yeast. Nature 440:940–943

20. Girhard M, Machida K, Itoh M, Schmid RD, Arisawa A, Urlacher VB (2009) Regioselective biooxidation of (+)-valencene by recombinant *E. coli* expressing CYP109B1 from *Bacillus subtilis* in a two-liquid-phase system. Microb Cell Fact 8:36

21. Gavira C, Höfer R, Lesot A, Lambert F, Zucca J, Werck-Reichhart D (2013) Challenges and pitfalls of P450-dependent (+)-valencene bioconversion by *Saccharomyces cerevisiae*. Metab Eng 18:25–35

22. Cankar K, van Houwelingen AMML, Bosch HJ, Sonke T, Bouwmeester H, Beekwilder MJ (2011) A chicory cytochrome P450 mono-

oxygenase CYP71AV8 for the oxidation of (+)-valencene. FEBS Lett 585:178–182

23. Sherman F (1991) Getting started with yeast. Methods Enzymol 194:3–21

24. Bergman LW (2001) Growth and maintenance of yeast. Methods Mol Biol 177:9–14

25. Liu J, Zhu Y, Du G, Zhou J, Chen J (2013) Exogenous ergosterol protects *Saccharomyces cerevisiae* from D-limonene stress. J Appl Microbiol 114:482–491

26. Brennan TCR, Turner CD, Krömer JO, Nielsen LK (2012) Alleviating monoterpene toxicity using a two-phase extractive fermentation for the bioproduction of jet fuel mixtures in *Saccharomyces cerevisiae*. Biotechnol Bioeng 109:2513–2522

Chapter 9

Purification of a Recombinant Polyhistidine-Tagged Glucosyltransferase Using Immobilized Metal-Affinity Chromatography (IMAC)

Fernanda de Costa, Carla J.S. Barber, Pareshkumar T. Pujara, Darwin W. Reed, and Patrick S. Covello

Abstract

Short peptide tags genetically fused to recombinant proteins have been widely used to facilitate detection or purification without the need to develop specific procedures. In general, an ideal affinity tag would allow the efficient purification of tagged proteins in high yield, without affecting its function. Here, we describe the purification steps to purify a recombinant polyhistidine-tagged glucosyltransferase from *Centella asiatica* using immobilized metal affinity chromatography.

Key words IMAC, His-tagged protein, Glucosyltransferase, Recombinant, Purification

1 Introduction

Immobilized metal affinity chromatography (IMAC) represents a viable and an efficient technique for the purification of proteins. In principle, IMAC can be used to purify proteins with natural surface-exposed mono- or oligo-histidine residues. More commonly, these histidine residues are genetically engineered "His-tags" of recombinant proteins [1].

The concept of IMAC is based on the known affinity of transition metal ions such as Zn^{2+}, Cu^{2+}, Ni^{2+}, and Co^{2+} towards amino acids like histidine and cysteine in aqueous solutions. One of the most important applications of this technique is the purification of recombinant proteins expressed in fusion with a sequence containing six or more histidine residues, commonly named as the polyhistidine-tag or his-tag [2].

Polyhistidine-tags present strongest interaction with immobilized metal ion matrices, since electron donor groups on the histidine imidazole ring preferably form coordination bonds with immobilized transition metal ions. Peptides containing sequences of

Arthur Germano Fett-Neto (ed.), *Biotechnology of Plant Secondary Metabolism: Methods and Protocols*, Methods in Molecular Biology, vol. 1405, DOI 10.1007/978-1-4939-3393-8_9, © Springer Science+Business Media New York 2016

consecutive histidine residues are efficiently held on IMAC column matrices. Such peptides can then be efficiently eluted by adding free imidazole or reducing the pH of the column buffer [3].

Purification of polyhistidine affinity-tagged proteins has been mainly performed using matrices containing divalent nickel or cobalt, coupled to a solid support resin (such as agarose or sepharose) functionalized with a chelator, such as iminodiacetic acid (Ni-IDA) and nitrilotriacetic acid (Ni-NTA) for nickel and carboxymethyl aspartate (Co-CMA) for cobalt. These bind metal ions through four coordination sites while leaving two of the transition metal coordination sites exposed to interact with histidine residues in the affinity tag [4]. Cobalt resins, compared to those containing nickel, have lower affinity for the polyhistidine affinity tag, resulting in elution of the tagged proteins under mild conditions. It has also been reported that Co^{2+} exhibits less nonspecific protein binding than the Ni^{2+}, resulting in higher elution product purity and minimal metal contamination [3].

An ideal affinity tag should enable effective but not overly strong binding, and allow elution of the desired protein under mild, nondestructive conditions. In the case of recombinant *E. coli*, many host proteins strongly adhere to the IMAC matrices, especially when charged with Cu^{2+} or Ni^{2+} ions, and are eluted with the target proteins [5]. Other limitations associated with affinity tags are poor capture of proteins expressed at low levels and protein loss during washing of the purification resin.

Protein purification is a complex process and a wide range of variables during purification can affect its success or failure. Considering the efforts to overcome the limitations cited above, this chapter presents purification steps that allowed us to reach a purified recombinant glucosyltransferase from the medicinal plant species *Centella asiatica*.

2 Materials

Prepare all solutions using distilled water and analytical grade reagents. Prepare and store all reagents at 4 °C (unless indicated otherwise). 1 L of sterile water should be put at 4 °C 2 days before experiment.

2.1 Protein Extraction Components

1. B-PER bacterial protein extraction reagent (Thermo Scientific, Rockford, IL, USA).

2. Stock solutions: Prepare a stock solution of 1 M Tris–HCl pH 7.0. Add about 900 mL water to a flask or beaker. Weigh 24.2 g of Tris base and transfer to the cylinder. Add water to a volume of 900 mL. Mix and adjust to pH to 7.0 with the appropriate volume of concentrated HCl. Bring final volume to 1 L with distilled water. Autoclave and store at room

temperature. Prepare a stock solution of 5 M NaCl solution in water. Weigh 146.1 g sodium chloride acid and prepare a 500 mL solution with water. Sterilize solution by autoclaving. Store at room temperature. Prepare a stock solution of 4 M imidazole solution in water. Weigh 27.3 g of imidazole and prepare a 100 mL solution with water. Sterilize solution by autoclaving and store in a bottle wrapped with aluminum foil at room temperature.

3. Equilibration/binding buffer: Prepare 50 mL of the equilibration/binding buffer inside a laminar flow hood on the day of the experiment (recommended) and keep it on ice. Mix 10 mL of the stock solution of 1 M Tris–HCl pH 7.0, 8 mL of the stock solution of 5 M NaCl solution, and 125 µL of the stock solution of 4 M imidazole solution in a 50 mL bottle and complete the volume with distilled water previously autoclaved and refrigerated. The final concentration of the binding solution will be 200 mM Tris–HCl, 800 mM NaCl, and 10 mM imidazole. Wrap the bottle with aluminum foil and keep it on ice until use.

2.2 Cobalt IMAC

1. HisPur™ cobalt resin (Thermo Scientific, Rockford, IL, USA).

2. Washing buffer: In a laminar flow hood, mix 20 mL of equilibration/binding buffer with 20 mL of sterile distilled water. Wrap the bottle with aluminum foil and keep it on ice until use.

3. Elution buffer: In the hood, mix 20 mL of equilibration/binding buffer, 20 mL sterile distilled water, and 1.875 mL of 4 M imidazole solution, resulting in a concentration of 150 mM. Wrap the bottle with aluminum foil, as imidazole is light sensitive, and keep it on ice until use.

4. LoBind microcentrifuge tubes (1.5 or 2 mL; Life Science Biotechnology, Hamburg, Germany).

3   Methods

3.1 Extraction of Protein from Bacteria

1. Pellet bacterial cells by centrifugation at $5000 \times g$ for 10 min.

2. Weigh pellet and add appropriately 4 mL of B-PER bacterial protein extraction buffer per gram of cell pellet.

3. Pipette the suspension up and down until it is homogeneous.

4. Transfer the lysate into multiple protein LoBind microcentrifuge tubes (1.5 or 2 mL), taking care not to fill them more than half full.

5. Incubate for 15 min on ice.

6. Alternate 2 min of sonication (on ice) and 2-min rest on ice. Repeat this procedure at least three times until cells are broken (viscous appearance).

7. Centrifuge lysate at $15,000 \times g$ for 8 min and separate the supernatant to new microtubes. Discard the pellet (*see* **Note 1**).

8. Add 1 volume of equilibration/binding buffer to each microtube to give the protein extract (*see* **Note 2**).

3.2 Preparation of the Cobalt Resin for IMAC Purification

The following procedures can be carried out on an open bench and must be on ice, unless otherwise stated. This first procedure can be done at room temperature.

1. Add an appropriate amount of cobalt resin to a tube. The recommended proportion of bacterial cell pellet (g) and resin (mL) is 2:1. For example, for 1 g of bacterial cell pellet, use 0.5 mL of resin.

   For optimal preparation, add maximum 250 μL resin in each 1.5 mL microcentrifuge tube.

2. Centrifuge at $700 \times g$ for 2 min and discard supernatant.

3. Mix pellet of resin with 2 volumes of washing buffer.

4. Centrifuge at $700 \times g$ for 2 min and discard supernatant.

5. Repeat **steps 3** and **4**.

3.3 Binding of Protein Sample

1. Add the resin to the microtube(s) containing the protein extract in equal amounts.

2. Rotate samples for 1 h at 4 °C.

3. Centrifuge at $700 \times g$ for 2 min at 4 °C.

4. Remove supernatant as the "unbound fraction" and this can be collected for protein quantification purposes (*see* **Notes 3–6**). The unbound fraction should be stored at 4 °C.

5. Wash resin adding 1 mL of washing buffer per microtube.

6. Centrifuge at $700 \times g$ for 2 min at 4 °C.

7. Remove supernatant as the "washing fraction." An aliquot of each fraction can be retained for protein quantification by UV spectrometry at 280 nm and to run SDS-PAGE. The washing fraction aliquot should be stored at 4 °C.

8. Repeat **steps 5–7** four times.

9. Measure optical density at 280 nm of each aliquot of the washing fraction to determine the protein concentration and success of washing procedure (*see* **Note 7**). Repeat **steps 5–7** until the signal of the eluate at 280 nm returns to near baseline.

3.4 Elution of Protein Sample

1. Elute protein from resin adding 1 mL of equilibration/binding buffer per microtube.

2. Centrifuge at $700 \times g$ for 2 min at 4 °C.

3. Remove supernatant as the "elution fraction." An aliquot of each fraction must be retained for protein quantification by

UV spectroscopy at 280 nm. Each elution fraction aliquot should be stored at 4 °C. The bulk of the protein of interest is generally found in the second and third elution fractions.

4. Measure optical density at 280 nm of each aliquot of the elution fraction to determine the success of elution procedure (*see* **Notes 8–11**). Repeat **steps 1–3** until the absorbance of 280 nm is near baseline and the elution of the protein of interest is complete.

3.5 Desalting and Concentration of Protein Samples

Desalting is a simple and fast method to remove low-molecular-weight contaminants and transfer the sample into the desired buffer in a single step. One possible method to achieve it is with the use of centrifugal filters. We used Amicon Ultra Centrifugal Filters Ultracel (EMD Millipore, Billerica, MA, USA) and the product line includes five different MW cutoffs, and the cutoff must be chosen taking into account the molecular weight of the desired protein.

1. Eluted fractions should be subsequently pooled together into a single centrifuge tube. The same procedure should be followed for the washing fractions.

2. Centrifuge the elution, washing, and unbound fraction(s) at $14,000 \times g$ for 15 min and discard the residual pellet.

3. Transfer each fraction to Amicon Ultra centrifugal filters of 10 or 30 kDa cutoff (depending on protein relative molecular mass) separately to dilute/desalt until the desired concentration of salts is reached. Normally, dilution of 1:5 (sample/desired buffer) is adequate and can be repeated as many times as necessary.

4. Centrifuge each sample at $1500 \times g$ at 4 °C for 10–40 min (variable depending on sample).

5. After completion of the desalting procedure, the recovery of the concentrated solute can be done by inserting a pipettor into the bottom of the filter device and withdrawing the sample using a side-to-side sweeping motion to ensure total recovery. The filtrate can be stored in the centrifuge tube.

6. Check the different fractions for proteins (e.g., by SDS-PAGE and/or Western blotting).

4 Notes

1. Bacterial cells can be re-extracted to optimize yield repeating the same procedure as Subheading 3.1.

2. Sometimes overexpressed proteins are sequestered in inclusion bodies inside the cell, decreasing the yield obtained after purification. Inclusion bodies of His-tagged proteins can be solubilized

in 8 M urea and 6 M guanidine and purified with the cobalt resin [3], but a denaturant must be added to buffers so the protein remains soluble throughout the procedure.

3. The low concentration of imidazole in the equilibration/binding buffer prevents the nonspecific binding; nevertheless, it allows binding of the His-tagged fusion proteins. Normally, concentration should be 10–20 mM. However, if His-tagged protein does not bind, imidazole concentration should be reduced to 1–5 mM.

4. Binding should be performed under conditions for which background proteins (contaminants) cannot compete for the binding sites of the cobalt resin. Sometimes the solution is to slightly reduce the pH and reduce imidazole concentration. For *C. asiatica* glucosyltransferase, the reduction of pH from 7.5 to 7.0 significantly increases the yield and purity of purified protein in the elution buffer.

5. All buffers should have sufficient ionic strength to prevent nonspecific binding. The minimum concentration of buffers should be around 300 mM and maximum around 2 M.

6. If the protein does not bind to the resin: sequence the DNA clone again to check if the reading frame is correct or check for premature termination sites. Perform an ELISA or Western blot using an antibody against the His-tag to make sure that the His-tag is present.

7. If the protein elutes with washing buffer, the solution can be used to reduce the concentration of imidazole or increase the pH slightly. Ensure that reducing or chelating agents are not present.

8. If the protein precipitates during purification, the temperature could be too low. Perform the purification at room temperature or try adding some solubilization reagents such as 0.1 % Triton X-100 or Tween-20, up to 20 mM β-mercaptoethanol, up to 2 M NaCl, or stabilizing cofactors such as Mg^{2+}.

9. If protein does not elute from the resin, elute with a pH gradient or with an imidazole gradient to determine the optimum elution conditions.

10. The inclusion of 10–20 mM imidazole in the equilibration/binding and in washing buffer should reduce/eliminate the elution of contaminants together with the protein. If contaminants may be covalently linked by disulfide bridges to the tagged protein, add β-mercaptoethanol to a maximum of 20 mM to reduce disulfide bonds. Wash resin additional times or modify imidazole concentration and/or pH of the equilibration/binding and washing buffer.

11. The cobalt resin may be used up to three times without affecting protein yield or purity. For His-Pur cobalt resin, regeneration procedures include washing with 2-(N-morpholino)ethane sulfonic acid (MES) buffer followed by washing with ultrapure water (see the manufacturer's instructions for details) and should be performed between each use of the product to eliminate possible contaminants, such as nonspecifically adsorbed proteins in the column. Resin should be stored as 50 % slurry in 20 % ethanol at 4 °C. Preferably, resin should be used for a specific fusion protein to prevent cross-contamination.

References

1. Gaberc-Porekar V, Menart V (2001) Perspectives of immobilized-metal affinity chromatography. J Biochem Biophys Methods 49:335–360

2. Block H, Maertens B, Spriestersbach A, Brinker N, Kubicek J, Fabis R, Labhan J, Schäfer F (2009) Immobilized-metal affinity chromatography (IMAC): a review. Methods Enzymol 463:439–473

3. Bornhorst JA, Falke JJ (2000) Purification of proteins using polyhistidine affinity tags. Methods Enzymol 326:245–254

4. Arnau J, Lauritzen C, Petersen GE, Pedersen J (2006) Current strategies for the use of affinity tags and tag removal for the purification of recombinant proteins. Protein Expr Purif 48:1–13

5. Arnold FH (1991) Metal-affinity separations: a new dimension in protein processing. Biotechnology 9:151–156

Chapter 10

A Western Blot Protocol for Detection of Proteins Heterologously Expressed in *Xenopus laevis* Oocytes

Morten Egevang Jørgensen, Hussam Hassan Nour-Eldin, and Barbara Ann Halkier

Abstract

Oocytes of the African clawed frog, *Xenopus laevis*, are often used for expression and biochemical characterization of transporter proteins as the oocytes are particularly suitable for uptake assays and electrophysiological recordings. Assessment of the expression level of expressed transporters at the individual oocyte level is often desirable when comparing properties of wild type and mutant transporters. However, a large content of yolk platelets in the oocyte cytoplasm makes this a challenging task. Here we report a method for fast and easy, semiquantitative Western blot analysis of proteins heterologously expressed in *Xenopus* oocytes.

Key words Transport protein quantification, Single oocyte Western blot, *Xenopus* oocytes

1 Introduction

Xenopus laevis is a tractable model organism that is widely used across diverse areas of research. *Xenopus* oocytes are especially well suited to investigate the biochemical properties of heterologously expressed channels and transporters. An important feature of the *Xenopus* oocyte is the low background activity of endogenous transporters and channels, which ensures a high signal-to-noise ratio for electrophysiological measurements and uptake assays. *Xenopus* oocytes have been used successfully to express and characterize membrane proteins from plants [1, 2], animals [3, 4], and microbes [5, 6]. When characterizing transport properties of membrane proteins expressed in *Xenopus* oocytes it is often desirable to compare protein expression levels between individual oocytes. Western blotting is a widely used semiquantitative technique for protein detection that uses antibodies to detect specific proteins that have been separated from one another by gel electrophoresis.

Arthur Germano Fett-Neto (ed.), *Biotechnology of Plant Secondary Metabolism: Methods and Protocols*, Methods in Molecular Biology, vol. 1405, DOI 10.1007/978-1-4939-3393-8_10, © Springer Science+Business Media New York 2016

Xenopus oocytes accumulate very large quantities of insoluble lipoproteins (up to 80 % of total oocyte protein) in granules referred to as yolk platelets [7]. This large lipoprotein content in the oocyte cytoplasm interferes with protein separation by sodium dodecyl sulfate–polyacrylamide gel electrophoresis (SDS-PAGE) and necessitates an initial precipitation of yolk platelets. Here, we report a simple method for fast and reliable semiquantitative detection of proteins heterologously expressed in *Xenopus oocytes* by Western blot analysis.

In Western blotting experiments proteins are separated by SDS-PAGE and transferred to a nitrocellulose or polyvinylidene fluoride (PVDF) membrane. The protein of interest is subsequently visualized by immunodetection with antibodies. Application of primary protein- or tag-specific antibody followed by washing leads to primary antibody only being bound to the protein of interest. The secondary antibody conjugated to an enzyme or molecule (e.g., horseradish peroxidase (HRP) or fluorochromes) allows the detection of the protein of interest with band intensity corresponding semiquantitatively to the amount of protein.

Instructions on how to perform gel separation and protein transfer from gel to PVDF membrane vary depending on the manufacturer. A detailed protocol should be consulted prior to experiments and can be found online on the manufacturer's homepage.

2 Materials

2.1 Preparation and Microinjection of cRNA Encoding Protein of Interest

The coding sequence (CDS) of your gene of interest is fused to an epitope tag (this protocol uses Human influenza hemagglutinin (HA) tag as an example, *see* **Note 1**) and cloned into a *Xenopus laevis* expression vector (e.g., the USER compatible pNB1u [8], https://www.addgene.org/62940/, *see* Fig. 1). For successful expression the vector must contain 5′ and 3′ UTRs of the endogenous *Xenopus* β-globin gene as well as a promoter upstream of the 5′ UTR, which is compatible with commercially available recombinantRNA polymerases (e.g., T7, T3, and SP6) [9].

1. T7 forward primer: 5′-TAATACGACTCACTATAGGG-3′.

2. SP6 reverse primer: 5′-ATTTAGGTGACACTATAG-3′.

3. PCR purification kit: QIAquick PCR purification kit.

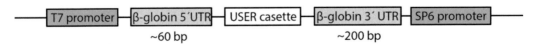

Fig. 1 Schematic drawing of the part of the *Xenopus* expression vector that is PCR-amplified and in vitro-transcribed. This part contains promoter, untranslated regions (UTRs), and USER cassette (into which gene of interest is cloned)

4. NanoDrop (or equivalent): Thermo Scientific.

5. RNAse decontamination solution (we use RNAseZap, Life technologies).

6. Pipette filtertips (nuclease free).

7. RNA in vitro transcription kit (we use mMESSAGE mMA-CHINE® T7 ULTRA Kit from Ambion®). Kit includes lithium chloride and nuclease-free H_2O for precipitation.

8. Nano-injector: Drummond NANOJECT II.

9. 1× Kulori buffer pH (7.4): 90 mM NaCl, 1 mM KCl, 1 mM $MgCl_2$, 1 mM $CaCl_2$, 5 mM HEPES, adjust to pH 7.4 with TRIS.

   • *Xenopus laevis* oocytes (can be purchased ready to inject from http://ecocyte-us.com/).

10. Ice-cold 70 % ethanol.

11. 1.5 mL Eppendorf tubes.

12. Tabletop centrifuge for 1.5 mL Eppendorf tubes.

2.2 Membrane and Soluble Protein Extraction from Oocytes

1. Homogenization buffer for protein extraction as described in [10]: 20 mM Tris–HCl (pH 7.6), 0.1 M NaCl, 1 % Triton X-100.

2. Roche protease inhibitor cocktail tablets.

2.3 SDS-PAGE and Western Blotting

1. 10–20 % Tris–HCl gels (we use Precast Criterion gels from Bio-Rad).

2. Protein size standard (we use Precision Plus Protein standard from Bio-Rad).

3. Protein electrophoresis and blotting apparatus (we use Criterion system from Bio-Rad).

4. Primary antibody 1 (e.g., rabbit-anti-HA).

5. Primary antibody 2 (e.g., anti-β actin antibody from Abcam, ab8227—loading control).

6. Secondary antibody (e.g., anti-rabbit antibody conjugated to horseradish peroxidase (HRP)).

7. 5× SDS sample buffer: 60 mM Tris–Cl pH 6.8, 2 % SDS, 10 % glycerol, 5 % β-mercaptoethanol, 0.01 % bromophenol blue.

8. Phosphate buffered saline (PBS, pH 7.4): 137 mM NaCl, 2.7 mM KCl, 10 mM Na_2HPO_4, 1.8 mM KH_2PO_4 (pH can be adjusted with HCl or NaOH).

9. PBS–Tween: PBS + Tween 20 (0.05 %).

10. PVDF membrane.

11. 100 % methanol.

12. Skimmed milk powder.

13. Chemiluminescent substrate for HRP (we use SuperSignal West Pico Chemiluminescent substrate from Thermo Scientific).

14. Imaging system to develop chemiluminescent blots (we use a ChemiDoc from UVP Bioimaging Systems).

15. Computer with ImageJ, an open access image analysis software (http://imagej.nih.gov/ij/).

3 Methods

3.1 Protein Expression in Xenopus Oocytes

1. The CDS of your protein of interest fused to an epitope tag, e.g., HA tag (*see* **Note 2**) is cloned into a *Xenopus* expression vector (e.g., pNB1, *see* Fig. 1 and https://www.addgene.org/62940/).

2. If pNB1 is used as expression vector, then PCR amplification of template with T7 and SP6 primers will yield PCR amplicons including the 5′ and 3′ *Xenopus* UTRs (*see* **Note 3**).

3. Purify PCR products using the QIAquick PCR Purification Kit following the manufacturer's guidelines (*see* **Note 4**).

4. Check DNA concentration by NanoDrop (or equivalent method). A concentration of 100 ng/μL or higher is needed for the subsequent in vitro transcription reaction.

5. Particular care has to be taken when working with RNA (including cRNA). It is of utmost importance that lab bench, pipettes and gloves are RNAse-free and wiped with RNAseZap (or equivalent). Pipette tips should be filter-tipped and RNAse-free.

6. In vitro transcription of cRNA is carried out on the purified PCR product (concentration ≥ 100 ng/μL) using the mMESSAGE mMACHINE Kit (*see* **Note 5**).

7. Precipitate cRNA by adding 60 μL 7.5 M lithium chloride and 60 μL nuclease-free H_2O to the reaction mixture and incubate at –20 °C for ≥ 2 h (*see* **Note 6**).

8. Pellet cRNA by centrifugation at $20,000 \times g$ for 15 min at 4 °C (*see* **Note 7**).

9. Wash resulting pellet in 1 mL ice-cold 70 % ethanol.

10. Centrifuge at $20,000 \times g$ for 5 min at 4 °C and remove supernatant.

11. Air-dry pellet for 2 min and resuspend in nuclease-free H_2O (*see* **Note 8**).

12. Determine cRNA concentration and quality by UV absorbance on a NanoDrop™ (Thermo Scientific) spectrophotometer (*see* **Note 9**).

13. Dilute all cRNA samples to 500 ng/μL prior to injection.

14. Defolliculated oocytes are purchased from Ecocyte Bioscience (http://ecocyte-us.com/). Alternatively, *see* ref. 11 for frog operation and oocyte isolation.

15. Inject 50 nL cRNA (500 ng/µL) into *Xenopus* oocytes using the Drummond NANOJECT II.

16. Incubate oocytes at 17 °C in Kulori pH 7.4 (change daily) for 1–7 days (*see* **Note 10**).

17. Save oocytes from Two Electrode Voltage Clamp (TEVC) electrophysiology measurements or representative oocytes from same batch of injected oocytes when assaying by LCMS-based uptake analysis (*see* **Note 11**).

3.2 Protein Extraction from Oocytes

1. Precipitation by centrifugation of insoluble yolk platelets is the crucial part of this method. In this step we solubilize membranes by addition of detergent (1 % Triton X-100) and subsequently spin down cell debris and yolk platelets.

2. Add one oocyte to a 1.5 mL Eppendorf tube (*see* **Note 12**), remove excess Kulori buffer and add 100 µL homogenization buffer with protease inhibitor cocktail.

3. Homogenize oocyte by pipetting.

4. Leave homogenate for 20 min at 4 °C to solubilize membrane proteins.

5. Centrifuge for 2 min at $10,000 \times g$ to precipitate yolk platelets and cell debris.

6. Transfer supernatant containing solubilized membrane proteins as well as soluble proteins to new tube (*see* **Note 13**).

7. Add 25 µL of 5× SDS sample buffer to supernatant and mix by pipetting (*see* **Note 14**).

3.3 SDS-PAGE and Western Blotting

1. Prepare 10–20 % Tris–HCl gels, electrophoresis chamber, and buffers according to the manufacturer's guidelines.

2. Load 10 µL of sample (mixed with SDS sample buffer) and a lane of protein size standard (*see* **Note 15**).

3. Separate proteins according to manufacturer's guidelines.

4. Prepare protein transfer apparatus according to the manufacturer's guidelines.

5. Transfer protein following the manufacturer's guidelines.

6. Block PVDF membrane overnight in PBS with 5 % skimmed milk at 4 °C and mild agitation.

7. Wash blocked membrane in PBS + 0.05 % Tween 20 for 2×5 min followed by 1×5 min in PBS (*see* **Note 16**).

3.4 Incubation with Primary and Secondary Antibodies

1. The optimal incubation time of the PVDF membrane together with primary and secondary antibodies depends on the antibodies, the antibody dilution and the amount of expressed proteins. In consequence, this part of the protocol requires individual optimization and the values listed here are merely for guidance.

2. To control for sample loading we include a primary antibody against the oocytes β-actin by multiplexing primary antibodies (*see* **Note 17**). This enables normalization of protein expression to the amount of loaded sample.

3. Incubate PVDF membrane with primary antibodies for 2–24 h followed by 3× wash for 5–10 min each in PBS + 0.05 % Tween 20 (*see* **Notes 17** and **18**).

4. Incubate the PVDF membrane in secondary antibodies conjugated to HRP for 2 h and wash twice for 5–10 min in PBS + 0.05 % Tween 20 and one time in PBS (*see* **Note 19**).

5. Remove PVDF membrane from washing solution and add SuperSignal West Pico chemiluminescent substrate and measure chemiluminescence on a chemiluminescence detector (e.g., ChemiDoc from UVP Bioimaging Systems).

3.5 Densitometric Quantification of Western Blot Bands

1. Densitometry is frequently used for semiquantitative analysis of the intensity of Western blot bands (*see* **Note 20**).

2. We use ImageJ, an open access image analysis software (http://imagej.nih.gov/ij/) and the standard "gel analysis" tool according to the developer's guidelines (see http://imagej.nih.gov/ij/docs/guide/user-guide.pdf section 30.13).

4   Notes

1. We have successfully used both HA and YFP tags as epitopes.

2. The HA tag is short and can be included in a cloning primer for easy fusion to the gene of interest. The sequence of the HA tag is [12]:

   (a) Protein sequence: YPYDVPDYA.

   (b) DNA sequence: tacccatacgatgttccagattacgct.

3. 5′and 3′ UTRs from the *Xenopus* β-globin gene are essential for efficient and high translation of heterologously expressed protein. Also, it is important that an optimized PCR reaction is used so only a single PCR band (and no spurious bands) is generated in the PCR amplification.

4. Ethanol inhibits the cRNA in vitro transcription reaction. In consequence, it is essential for high-quality and good yield of cRNA that the PCR product is completely free of ethanol. To ensure this, add an additional spin step to remove all ethanol from the wash step.

5. To save costs, a noncommercial in vitro transcription method has been described and is routinely used in our lab [8]. However, the capping efficiency is lower and consequently expression is lower using the homemade kit. We typically use cRNA produced via the homemade kit for LCMS-based uptake experiments whereas we use the commercial kit for making cRNA destined for electrophysiological experiments in single oocytes to obtain maximal protein expression.

6. We commonly perform overnight precipitation to avoid unnecessary loss of cRNA. Be careful not to suck up the pellet when removing supernatant.

7. Be careful when removing supernatant. The pellet can be difficult to see and should not be touched.

8. Do not over-dry pellet as this greatly complicates resuspension. In case precipitated cRNA is insoluble, we recommend freezing the samples at −20 °C and then thawing it followed by careful pipetting.

9. A 260/280 absorbance ratio of ~2 is an indication of pure cRNA. We aim for a cRNA concentration above 1000 ng/μL when following the manufacturer's guidelines for in vitro transcription.

10. Incubation time for proper expression varies dependent on the gene being expressed (Fig. 2).

11. Depending on the type of experiment we quantify protein levels in a single oocyte or as the average of several oocytes. For Two Electrode Voltage Clamp (TEVC) electrophysiology experiments, we quantify expressed protein level in single oocytes. For the less sensitive LCMS-based measurements, we typically use five oocytes per uptake assay and quantify the

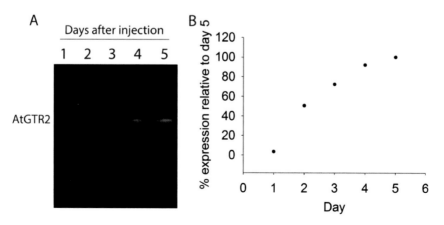

Fig. 2 Detection and relative quantification of HA-tagged AtGTR2 transporter from *Arabidopsis thaliana* expressed in *Xenopus* oocytes for 1–5 days. (**a**) Western blot on protein extracts from single oocytes at day 1–5 after injection. (**b**) Densitometric analysis of gel band intensity from (**a**)

expression level as the average of five oocytes. It is important that the ratio of oocyte to homogenization buffer remains 1:100 µL.

12. Oocytes can be stored at −20 °C at this time point.

13. A thin, white, oil-like layer can form above the aqueous phase. Avoid pipetting this solution (it does not interfere).

14. Do not boil samples as this is not necessary and may lead to loss of sample due to aggregation.

15. The expression level influences the amount of sample to be loaded. If Western blot bands are weak try loading 20 µL sample.

16. Washing regime is dependent on the antibodies used and this should be taken into account when optimizing blotting protocol.

17. A low primary antibody concentration in combination with a prolonged incubation time is recommended to ensure specific binding (see notes for concentrations). Test a range of dilutions the first time you use your primary antibody (typically 1:100–1:5000).

18. We include an anti β-actin antibody to monitor sample loading. As for the other primary antibodies, a dilution series should be tested prior to actual experiments.

19. The concentration of secondary antibodies can generally be higher than the primary antibody concentration (1:1000–1:5000). We recommend using HRP conjugation over alkaline phosphatase conjugation because of the higher sensitivity of HRP.

20. Several subtle pitfalls exist when performing densitometric analysis of Western blot bands. We recommend the following literature as reading material [13].

References

1. Boorer KJ, Forde BG, Leigh RA, Miller AJ (1992) Functional expression of a plant plasma membrane transporter in *Xenopus* oocytes. FEBS Lett 302:166–168

2. Theodoulou FL, Miller AJ (1995) *Xenopus* oocytes as a heterologous expression system for plant proteins. Mol Biotechnol 3:101–115

3. Sumikawa K, Houghton M, Emtage JS, Richards BM, Barnard EA (1981) Active multi-subunit ACh receptor assembled by translation of heterologous mRNA in *Xenopus* oocytes. Nature 292:862–864

4. Sigel E (1990) Use of *Xenopus* oocytes for the functional expression of plasma membrane proteins. J Membr Biol 117:201–221

5. Calamita G, Bishai WR, Preston GM, Guggino WB, Agre P (1995) Molecular cloning and characterization of AqpZ, a water channel from *Escherichia coli.* J Biol Chem 270:29063–29066

6. Wahl R, Wippel K, Goos S, Kämper J, Sauer N (2010) A novel high-affinity sucrose transporter is required for virulence of the plant pathogen *Ustilago maydis.* PLoS Biol 8:e1000303

7. Redshaw MR, Follett BK (1971) The crystalline yolk-platelet proteins and their soluble plasma precursor in an amphibian, *Xenopus laevis.* Biochem J 124:759–766

8. Nour-Eldin HH, Norholm MHH, Halkier BA (2006) Screening for plant transporter func-

tion by expressing a normalized *Arabidopsis* full-length cDNA library in *Xenopus* oocytes. Plant Methods 2:17

9. Krieg PA, Melton DA (1984) Functional messenger RNAs are produced by SP6 in vitro transcription of cloned cDNAs. Nucleic Acids Res 12:7057–7070

10. Galili G, Altschuler Y, Ceriotti A (1995) Synthesis of plant proteins in heterologous systems: *Xenopus laevis* oocytes. In: David W, Galbraith DPB, Hans JB (eds) Method cell biol. Academic, San Diego, pp 497–517

11. Liu XS, Liu XJ (2006) Oocyte isolation and enucleation. Methods Mol Biol 322:31–41

12. Wilson IA, Niman HL, Houghten RA, Cherenson AR, Connolly ML, Lerner RA (1984) The structure of an antigenic determinant in a protein. Cell 37:767–778

13. Gassmann M, Grenacher B, Rohde B, Vogel J (2009) Quantifying Western blots: pitfalls of densitometry. Electrophoresis 30: 1845–1855

Chapter 11

Laser Capture Microdissection: Avoiding Bias in Analysis by Selecting Just What Matters

Márcia R. de Almeida and Martina V. Strömvik

Abstract

Laser capture microdissection (LCM) is a powerful technique for harvesting specific cells from a heterogeneous population. As each cell and tissue has its unique genetic, proteomic, and metabolic profile, the use of homogeneous samples is important for a better understanding of complex processes in both animal and plant systems. In case of plants, LCM is very suitable as the highly regular tissue organization and stable cell walls from these organisms enable visual identification of various cell types without staining of tissue sections, which can prevent some downstream analysis. Considering the applicability of LCM to any plant species, here we provide a step-by-step protocol for selecting specific cells or tissues through this technology.

Key words Cell-specific technique, Cryosections, Plant cells, Microgenomics, Microscopy

1   Introduction

The majority of biological processes is asymmetrically distributed in higher plants and is dependent of specific trends, activities, and interactions of subsets of specialized cells in determined locations [1, 2]. Because of this, the genetic, proteomic, and metabolic profiles of each cell are unique [3]. In the actual large-scale data era, the advances in data generation and analysis have helped the understanding of correlations between transcriptome, proteome, and metabolome profiles. However, at the same time, the complexity of several developmental processes, which are affected by both biotic and abiotic factors, highlights the necessity of more detailed analysis, to overcome the possible bias when analyzing samples composed by heterogeneous groups of cells [4]. Considering this, the use of techniques for isolation of specific tissue or cell types is a good strategy. Several techniques have been developed in the last decades and, although primarily used in medical research with animal cells, they have been successfully used in plant studies as well [5–10].

Arthur Germano Fett-Neto (ed.), *Biotechnology of Plant Secondary Metabolism: Methods and Protocols*, Methods in Molecular Biology, vol. 1405, DOI 10.1007/978-1-4939-3393-8_11, © Springer Science+Business Media New York 2016

The fluorescent activated cell sorting (FACS) allows the collection of specific cell types by tagging these cells with marker reporters, such as green fluorescent protein (GFP) or yellow fluorescent protein (YFP) [11]. Similarly, the isolation of nuclei targets in specific cell types (INTACT) allows the purification of the nucleus from the cells of interest through the expression of a nuclear targeting fusion protein (NTF) containing GFP and a biotin acceptor peptide [12]. These two techniques are very interesting but both have the limitation of needing transgenic plants, a difficult procedure still unsuccessful for many plant species and that requires a good knowledge of tissue specific promoters. INTACT has the additional limitation of being useful only for transcriptomic analysis.

Attempting to overcome the above-cited practical limitations, the laser microdissection (LM) has evolved as a powerful technique, which can be used for any species and for different purposes [1, 3]. In this technology, a particular microscope is used to select specific cells from fixed samples, based on morphology or histology [5]. LM can be divided in two major different techniques: Laser cutting and Laser capture microdissection (LCM). Several instruments are available on the market and all of them use one or both of these two strategies [1]. In laser cutting, the target cell is cut free from tissue section (non contact) by UV laser and collected by different methods such as laser pressure catapulting, ejection downward a collection tube or blotting onto an adhesive cap [1]. In LCM, the target cells are isolated from the tissue section by bonding them to a plastic film at sites activated by a near-IR laser, using a LCM cap as the collector device. The advantage of LCM when compared to other LM methods is that cells can be harvested onto the cap while their spatial relationships in the original section are preserved. With this, the inspection of LCM-isolated cells on the harvest cap is facilitated, allowing a detailed analysis of the harvested section after each event of collection. Furthermore, adjacent cells to the cells of interest are not damaged as in laser cutting, preserving them for future analysis [1]. The contact of the sample with the plastic film can be avoided by the use of membrane frame slides, when it is possible to mount tissue samples onto a membrane that separates the tissue from the film.

The low concentrations of RNA, proteins and metabolites obtained from LCM captured cells and tissues are still a challenge. However, the development of highly sensitive detection methods and effective protocols for amplification of the extracted products are collaborating for increasing the application of this technology in plants, helping the elucidation of complex regulatory networks occurring at the cellular level [2, 13]. Here we describe a general protocol for obtaining laser captured microdissected cells or tissues from plant samples. The present protocol was successfully applied in soybean [9] and *Eucalyptus* sp. (Fig. 1) and can be used for transcriptomic, proteomic, or metabolomic downstream studies. Cell

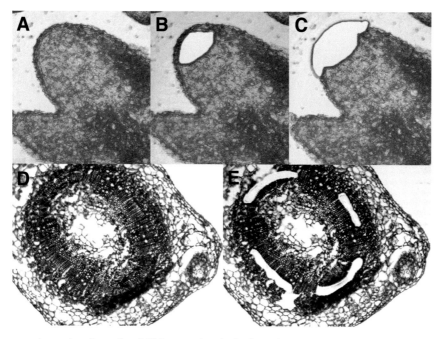

Fig. 1 Laser capture microdissection (LCM) procedure in Soybean (**a–c**) and *Eucalyptus globulus* (**d** and **e**). (**a**) Soybean shoot tip before LCM; (**b**) Soybean shoot tip after capture of shoot tip meristem tissue; (**c**) Soybean shoot tip after capture of shoot tip meristem and epidermis; (**d**) *E. globulus* stem cross section before LCM; (**e**) *E. globulus* stem cross section after capture of cambium cells

and tissue specific biochemical and molecular microanalyses are particularly useful in addressing aspects of plant secondary metabolism for often the expression of different metabolic steps of its biochemical pathways take place in distinct cell types and tissues.

2 Materials

Prepare all solutions using ultrapure water and analytical grade reagents. Ensure all utensils and glassware are sterile or have been previously cleaned and autoclaved (minimum of 20 min at 121 °C). Prepare and store all reagents at room temperature (unless indicated otherwise). Ensure all waste disposal regulations will be followed when disposing of waste materials. Ensure you are wearing powder-free gloves during all the procedures.

2.1 Sample Fixation and Infiltration

1. Vacuum pump or any other equipment to produce vacuum in the samples.

2. Glass Buchner flasks with proper cap.

3. Tweezers and blades.

4. Fixation solution: 75 % ethanol–25 % acetic acid solution v/v (*see* **Note 1**).

5. Shaker and ice bath or any ice appropriate container.

6. PBS Buffer: 137 mM NaCl, 8 mM Na_2HPO_4, 2.7 mM KCl, 1.5 mM KH_2PO_4, pH 7.3. Mix all the reagents in a glass beaker filled with about 1/3 of the final volume of ultrapure water. Set the final volume and adjust the pH. Ensure you are performing enough quantity of PBS Buffer for all of the following solutions. Store at 4 °C.

7. 10 % sucrose solution: Add 10 g of sucrose in 100 mL of PBS Buffer. Mix to dissolve.

8. 20 % sucrose solution: Add 20 g of sucrose in 100 mL of PBS Buffer. Mix to dissolve.

9. Sterile plastic centrifuge tubes (15 or 50 mL), depending on the size of your sample.

2.2 Sample Embedding in Cryomedium

1. Petri dishes.
2. Clean wipes.
3. Cryocassettes (*see* **Note 2**).
4. Cryomedium.
5. Liquid nitrogen.

2.3 Cryosectioning

1. Cryotome.
2. Tweezer and brushes.
3. Cryocassette.
4. UV treated PEN membrane frame slides (*see* **Note 3**).
5. Slides container or dry ice (*see* **Note 4**).

2.4 Dehydration

1. Ethanol solutions in ultrapure water: 70, 75, 95, and 100 % (*see* **Note 5**).
2. Xylene.
3. 200 and 1000 μL pipettes.
4. RNAse-free pipette tips.

2.5 Laser Capture Microdissection (LCM) Procedure

1. LCM instrument.
2. LCM caps.
3. RNAse-free glass support slides (*see* **Note 6**).

3 Methods

3.1 Sample Fixation and Infiltration

Proceed with the harvesting of the sample of interest. Use fresh samples. Any part of the plant can be used for the procedure. However, if you have a large sample, such as stems or leaves, you will probably want to cut it into small pieces to facilitate the fixation and infiltration processes. If necessary, gently wash the sample with autoclaved distilled water.

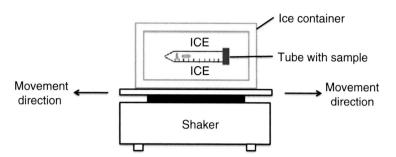

Fig. 2 Sample agitation during fixation and infiltration process. Procedure scheme for sample agitation on ice. Samples were incubated with fixation and infiltration solutions in 15 mL centrifuge tubes and kept immersed in ice as described in **step 2** from Subheading 3.1

1. Fill one Buchner flask with enough fixation solution to fully cover your sample. Put the sample in the flask and make sure all the sample is in contact with the solution. Close the flask and put it inside an ice bath or a container filled with ice. Connect the flask to the vacuum pump and leave the sample under vacuum for 15 min (*see* **Note 7**).

2. Put the sample in a sterile centrifuge tube and fill out the tube with fresh fixation solution. Place the tube (horizontally oriented) in a container filled with ice. Completely cover the tube with another layer of ice and close the container. Place the container in the shaker and set the speed to allow a gentle agitation (not too slow and not too fast) (*see* **Note 8**). Keep shaking overnight (Fig. 2).

3. Transfer the sample to a Buchner flask filled with 10 % sucrose solution. Place the flask inside an ice bath or a container filled with ice. Connect the flask to the vacuum pump and leave the sample under vacuum for 15 min.

4. Transfer the sample to a centrifuge tube filled with fresh 10 % sucrose solution and place the tube in the same ice container used for **step 2**. Let agitate for 30 min or until the samples are fully submerged.

5. Repeat **step 3** using the 20 % sucrose solution for 15 min.

6. Repeat **step 4** using the 20 % sucrose solution for 30 min or until the samples are fully submerged.

3.2 Sample Embedding in Cryomedium

1. Gently dry the samples with a clean paper towel.

2. Cool the cryocassette pouring a small amount of liquid nitrogen on it. Use a cold block as a basis for the cryocassette or perform the procedure in a large Styrofoam container, to avoid accidental contact with the liquid nitrogen.

3. Considering the size of the sample, pour an amount of cryomedium on the cold cryocassette to make a thick basis for your sample. Place the sample in the right orientation and immediately cover it with another thick layer of cryomedium (*see* **Note 9**).

4. Freeze immediately with liquid nitrogen (*see* **Note 10**). Detach from the cryocassette, wrap in aluminum foil and store at −80 °C until ready for sectioning (*see* **Note 11**).

3.3 Cryosectioning

1. Ensure the cryotome is clean and in adequate temperature, which is −21 °C (a range between −20 and −24 is ok). Keep blades, brushes, tweezers, and any other material you think necessary inside the cryotome while making sections.

2. Adjust the cryotome settings according to your sample and purpose (*see* **Note 12**).

3. Allow the "cryocassette" to cool by keeping it on the cryobar for ~1 min.

4. Put a drop of cryomedium on the cryocassette and put the sample (which is already embedded) in the desired orientation (*see* **Notes 13** and **14**).

5. Place the cryocassette back on the cryobar and wait until the whole assembly is frozen.

6. Fit the cryocassette containing the sample in the "sample head" of the cryotome and adjust the angle of the blade to start sectioning.

7. When the sample is in line with the blade, start slicing off sections of the medium until you reach the sample.

8. When the desired region of the tissue is reached either pick up the section with a frozen, fine paintbrush and place it on the slide, which is at room temperature, or place the flat side of the slide directly on the section and allow it to melt on top of the slide. As soon as you get the first section, keep the slide inside the cryotome until finished (*see* **Note 15**). When using PEN membrane frame slides, you should place the samples in the flat side of the slide (Fig. 3). Try to put the sections as much in the center of the slide as possible (*see* **Note 16**).

9. When you finish taking samples to the slides, you are ready to proceed to the dehydration step. If you cannot proceed immediately to the next step, keep the slides with the cryosections in a closed container at −80 °C until use (*see* **Note 17**).

3.4 Dehydration

Proceed with the dehydration directly on the slides. You can use immersion jars, Pasteur pipettes, or electronic pipettes for each solution (*see* **Note 18**).

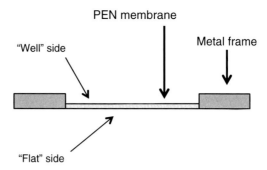

Fig. 3 Detail of a PEN membrane framed slide. Scheme showing the exact place where to put the cryosections

1. Ethanol 70 % (at −20 °C) for fixation—30 s to 1 min (*see* **Note 19**).

2. RNAse-free water to dissolve the cryomedium—30 s to 1 min.

3. Ethanol 75 %—30 s to 1 min.

4. Ethanol 95 %—30 s to 1 min.

5. Ethanol 100 %—30 s to 1 min.

6. Xylene—two times of 2 min each.

7. Air-dry and use immediately for LCM (*see* **Note 20**).

3.5 Laser Capture Microdissection (LCM) Procedure

Depending on the LCM instrument, the procedures regarding the orientation of slides and mode of cell capture can change. In this section we describe general procedures for instruments using infrared (IR) and ultraviolet (UV) lasers for capture and cut of cells or tissue of interest, respectively. Here we also focus on the cell collection methodologies using specific caps.

1. For correct orientation of the slide in the LCM instrument, you should use RNAse-free support glass slides (*see* **Note 6**) below the metal membrane frame slides (Fig. 4). Be sure your sections are completely dry before doing that.

2. Load slides and caps in their specific places in the instrument.

3. Adjust brightness and focus and search for your cells of interest. Once you find them, draw around the area of cells to be microdissected.

4. Place the cap so your cells of interest stay inside the coverage region of the cap.

5. Move to a clean part under the cap where there are no cells so you can manipulate the IR capture laser and UV cutting laser.

6. Once both lasers are settled you are ready to move to the drawn area to capture and cut the cells of interest (*see* **Notes 21** and **22**).

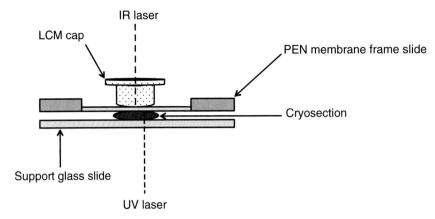

Fig. 4 Scheme showing the right orientation of a PEN membrane frame slide containing section on glass slide support. Oriented this way, the IR laser will allow the capture of the selected area by melting the thermoplastic transfer film in a region surrounding the selected cells or tissue of interest, so the cells are attached to the cap. Then the UV laser is used to cut the drawn area, completing the dissecting of the cells of interest. The glass slide support is necessary so the weight of the cap does not change the curvature of the film in the PEN membrane frame slide. The change of curvature can modify the angle with which the lasers hit the sample, potentially damaging the capture and cutting procedure

7. When finished sectioning, immediately place the cap containing your samples in a 0.5 mL microcentrifuge tube already filled with the appropriate buffer for further downstream processes.

8. Invert the cap–microcentrifuge tube assembly. Tap the bottom of the tube to ensure all buffer is in contact with the captured cells.

9. Incubate following specific recommendations from your reagent's supplier considering the purpose of your downstream process.

4   Notes

1. Wear gloves and a mask when preparing this solution. Perform the procedure preferably in a fume hood.

2. Cryocassettes are specific sample holders for using in cryotomes. If not available for using during sample embedding you can alternatively use petri dishes. Just be sure the petri dishes are not going to crack when in contact with liquid nitrogen.

3. Perform an UV treatment (15–30 min) in the PEN membrane frame slides to aid in the adhesion of the sections on the polymer membrane. This treatment should be done on the slides preferably just a few days before the procedure. The treatment can be performed in air flow cabinets with UV light or using

UV lamps. Be careful to avoid any direct contact of your eyes or skin with UV light.

4. While taking sections, keep the slides inside the cryotome or next to it on dry ice.

5. Stock in 70 % ethanol solution at −20 °C until use.

6. In an air flow cabinet, treat the glass support slides with a specific solution for elimination of RNAses and rinse with ultra-pure water. Air-dry before use.

7. Test the pressure before start. You know the vacuum is working when you see little bubbles coming out the sample.

8. Ensure the samples are moving inside the tube and are in contact with the solution. If a cold chamber is available, the ice is not necessary and you can replace the shaker with a rotator. Keeping the samples cold is very important, especially if you intend to extract RNA.

9. It is very important that the cryocassette is cold before starting the embedding procedure. Alternatively you can use small petri dishes as a basis for the embedding. In this case, cool the petri dish with liquid nitrogen and fill it with cryomedium. Place your sample and cover with more cryomedium. The only issue is that it is harder to take the samples out the petri dishes than from the cryocassettes. You can also use specific cryo-molds for this step.

10. You know the cryomedium is frozen when the color turns from clear to opaque.

11. Attempting to facilitate the sectioning, cut out the excess of cryomedium so you have flat sides in your sample. This is especially helpful for a good orientation of the sample to make cross sections.

12. The section thickness can vary depending on the purpose. Thin sections (1–10 μm) are better for visualization of structures and taking pictures. Thick sections (20–30 μm) are better for LCM, as you can take more cell layers at each time, but some structures can be difficult to detect in thick sections. We usually use 20–25 μm for LCM, but a previous test to define the better thickness for each case is advised.

13. If you prefer you can embed the sample just before sectioning, directly in the cryocassette that you are going to use in the cryotome. However, when you are working with more than one sample, it is better to proceed with the embedding right after the sample fixation and infiltration to avoid degradation of the samples and save time.

14. Ensure you are manipulating your embedded sample inside the cryotome and that your tweezers and blades are also cooled, to avoid melting the cryomedium.

15. Remember to keep all the slides containing sections inside the cryotome or next to it, in a container filled with dry ice, so the sections do not melt. This is very important if you intend to extract RNA.

16. Depending on your sample, you will probably need more than one slide with cryosections per sample. This is very important to ensure you will have enough samples to work within your downstream processes. It is a good idea to make a prior test to know how much cells or tissue section you need per sample.

17. Take care during the transport of the slides to the –80 °C to avoid melting the sections. This is very important to preserve RNA. We generally use a container filled with a small amount of liquid nitrogen. Then we take the slide box containing the slides off the cryotome and place it in contact with the liquid nitrogen for the transport. Avoid direct contact of the liquid nitrogen with the slides.

18. We prefer to use electronic pipettes during the dehydration steps since these are practical and reduce the chance of contamination if using sterilized tips. Be careful to avoid losing sections during the procedure by gently applying each solution to the slide. If you prefer to use immersion jars, ensure they are clean and autoclaved and that you have enough jars for each solution. However, the chances of losing sections using immersion jars do increase comparing to the use of pipettes.

19. To avoid melting of the cryosections, and risk RNA degradation, apply the 70 % ethanol (which was at –20 °C) inside the cryotome or just after taking out the sections from –80 °C. After this, proceed at room temperature. Make sure you spend at most 1 hour on microdissection in each slide to avoid increased degradation. High levels of humidity can also make the procedure more difficult.

20. If you see some white stains on the slide it means that the cryomedium was not completely dissolved, so you should start the whole procedure again.

21. If you use more than one cap to capture all the cells of interest, make sure to gather all the extracts to increase the concentration of your RNA, protein, or metabolite of interest. If using column based downstream procedures, pass the extracts from each cap from a same sample through the same column.

Acknowledgments

The authors would like to thank Haritika Majithia for providing soybean shoot tip pictures (Fig. 1). The development of the described protocol and preparation of this manuscript were supported by grants from the Natural Sciences and Engineering Research Council of Canada (NSERC), the Canadian Foundation for Innovation (CFI), the Dr. Louis G. Johnson Foundation, and the Commission for Improvement of Higher Education Personnel (CAPES, Brazil).

References

1. Nelson T, Tausta SL, Gandotra N, Liu T (2006) Laser microdissection of plant tissue: what you see is what you get. Annu Rev Plant Biol 57:181–201

2. Fang J, Schneider B (2014) Laser microdissection: a sample preparation technique for plant micrometabolic profiling. Phytochem Anal 25:307–313

3. Gautam V, Sarkar AK (2014) Laser assisted microdissection, an efficient technique to understand tissue specific gene expression patterns and functional genomics in plants. Mol Biotechnol 57:299–308

4. Mallick P, Kuster B (2010) Proteomics: a pragmatic perspective. Nat Biotechnol 28:695–709

5. Asano T, Masumura T, Kusano H, Kikuchi S, Kurita A, Shimada H, Kadowaki K (2002) Construction of a specialized cDNA library from plant cells isolated by laser capture microdissection: toward comprehensive analysis of the genes expressed in the rice phloem. Plant J 32:401–408

6. Spencer MW, Casson SA, Lindsey K (2007) Transcriptional profiling of the *Arabidopsis* embryo. Plant Physiol 143:924–940

7. Brooks L III, Strable J, Zhang X, Ohtsu K, Zhou R, Sarkar A, Hargreaves S, Elshire RJ, Eudy D, Pawlowska T, Ware D, Janick-Buckner D, Buckner B, Timmermans MCP, Schnable PS, Nettleton D, Scanlon MJ (2009) Microdissection of shoot meristem functional domains. PLoS Genet 5:e1000476

8. Abbott E, Hall D, Hamberger B, Bohlmann J (2010) Laser microdissection of conifer stem tissues: isolation and analysis of high quality RNA, terpene synthase enzyme activity and terpenoid metabolites from resin ducts and cambial zone tissue of white spruce (*Picea glauca*). BMC Plant Biol 10:106

9. Livingstone JM, Zolotarov Y, Strömvik MV (2011) Transcripts of soybean isoflavone 7-O-glucosyltransferase and hydroxyisoflavanone dehydratase gene homologues are at least as abundant as transcripts of their well known counterparts. Plant Physiol Biochem 49:1071–1075

10. Zhang X, Douglas RN, Strable J, Lee M, Buckner B, Janick-Buckner D, Schnable PS, Timmermans MCP, Scanlon MJ (2012) Punctate vascular expression1 is a novel maize gene required for leaf pattern formation that functions downstream of the trans-acting small interfering RNA pathway. Plant Physiol 159:1453–1462

11. Iyer-Pascuzzi AS, Benfey PN (2010) Fluorescence- activated cell sorting in plant developmental biology. Methods Mol Biol 655:313–319

12. Deal RB, Henikoff S (2011) The INTACT method for cell type–specific gene expression and chromatin profiling in *Arabidopsis thaliana*. Nat Protoc 6:56–68

13. Rogers ED, Jackson T, Moussaieff A, Aharoni A, Benfey PN (2013) Cell type-specific transcriptional profiling: implications for metabolite profiling. Plant J 70:5–17

Chapter 12

Analytical and Fluorimetric Methods for the Characterization of the Transmembrane Transport of Specialized Metabolites in Plants

Inês Carqueijeiro, Viviana Martins, Henrique Noronha, Hernâni Gerós, and Mariana Sottomayor

Abstract

The characterization of membrane transport of specialized metabolites is essential to understand their metabolic fluxes and to implement metabolic engineering strategies towards the production of increased levels of these valuable metabolites. Here, we describe a set of procedures to isolate tonoplast membranes, to check their purity and functionality, and to characterize their transport properties. Transport is assayed directly by HPLC analysis and quantification of the metabolites actively accumulated in the vesicles, and indirectly using the pH sensitive fluorescent probe ACMA (9-amino-6- chloro-2-methoxyacridine), when a proton antiport is involved.

Key words ABC transporters, ACMA, Alkaloids, BY-2 cells, *Catharanthus roseus*, H^+-antiport, MATE transporters, Tobacco, Tonoplast vesicles, Transmembrane transport, Transmembrane pH gradient, Uptake assay

1 Introduction

Plants exhibit a unique metabolic plasticity implicated in many aspects of their environmental success as sessile organisms. For humans, this specialized metabolism provides a wealth of natural products with innumerous applications as medicines, colorants, flavors, fragrances, cosmetics, etc. Currently, next generation sequencing is fuelling a burst in molecular knowledge on specialized pathways, driving intense research seeking to obtain increased yields of metabolites for commercial exploitation or improvement of plant defenses. However, in spite of intensive investigation, little is known about the transmembrane transport mechanisms involved in specialized pathways, even if they often involve organ/tissue/cell translocation events as well as subcellular compartmentation (e.g., nicotine, berberine, morphine, and vinblastine pathways) [1].

Arthur Germano Fett-Neto (ed.), *Biotechnology of Plant Secondary Metabolism: Methods and Protocols*, Methods in Molecular Biology, vol. 1405, DOI 10.1007/978-1-4939-3393-8_12, © Springer Science+Business Media New York 2016

Therefore, membrane transport of specialized metabolites is emerging as a newly developing research line, essential to the understanding of metabolic fluxes and to the implementation of metabolic engineering strategies aimed at increased levels of valuable metabolites.

Two major mechanisms have been implicated in the transmembrane transport of specialized metabolites in plants: primary transport mediated by ABC (ATP-binding cassette) transporters and secondary transport mediated by MATE (multidrug and toxic compound extrusion) transporters functioning as H^+-antiports [1, 2]. Both have been implicated in the transport of alkaloids and flavonoids, and ABC transporters have been further implicated in the transport of terpenoids, ABA, strigolactones and lipid and monolignol deposition in the extracellular surface [1, 2].

The isolation of pure and functional organelles or organelle-derived membrane vesicles is a prerequisite for successful transport studies, with the membrane sidedness being a key issue in the case of isolated vesicles. The determination of the radioactivity entrapped by the cells/organelles/membrane vesicles upon incubation with a radiolabeled substrate is a reliable and sensitive method to measure uptake [3], but its main drawback lies in the fact that the majority of alkaloids or other specialized metabolites are not commercially available as radioisotopes. On the other hand, in some cases, the transported solute may be quantified by highly sensitive analytical methods, including HPLC [4]. Uptake studies relying on electrophysiological measurements and specific fluorescent probes, sensitive for instance to the membrane potential or pH, are also widely reported [5]. Thus, the pH sensitive fluorescent probe ACMA (9-amino-6-chloro-2-methoxyacridine) has been frequently used to measure changes in vacuolar pH when a specific substrate crosses the tonoplast through a putative H^+/solute antiport system (*see* **Note 1**) [5]. Important issues when using membrane vesicles in transport experiments are the tightness and orientation of the membrane. While sealing may be optimized by fusing the biological membranes with liposomes, the manipulation of the orientation (right side out or inside-out) may be useful depending on both the type and localization of the transport system under study. Thus, the production of inverted vesicles from isolated plasma membranes (cytoplasmic side-out) may be achieved with Brij 58 [6]. This vesicle system is appropriate to study H^+-dependent antiport systems after exogenous addition of ATP that promotes the acidification of the vesicle lumen through the activation of P-ATPase.

A first good evidence for the involvement of a mediated transport system comes from the observation that the initial velocities of transport, over a specific range of substrate concentrations, follow a saturation kinetics, providing at the same time useful information about the kinetic parameters [7]. The simultaneous addition of the transported substrate and a putative competitor or inhibitor may allow the characterization of the specificity of the transport system [3, 8]. The thermodynamic aspects (energetics) involved in the substrate uptake may be experimentally evaluated by studying the H^+ (or Na^+) dependency of the transport system, by measuring, for instance, the uptake at different pH values, the effect of protonophores, or the involvement of a primary transport system, usually through the addition of ATP or PPi [5, 8].

Here, we describe the methodology used to characterize the vacuolar transport of the medicinal alkaloids from *Catharanthus roseus* [4], which was also used to study the transport of nicotine in tonoplast membranes from tobacco leaves and tobaccoBY-2 cells. This approach, which may be easily adapted to study the transport of other specialized metabolites in other plant models, involves the following steps:

1. Isolation of tonoplast vesicles.

2. Characterization of the purity of tonoplast vesicles by Western blot and assay of vacuolar (V)-H^+-ATPase hydrolytic activity by measurement of ATP hydrolysis in the presence of azide, a mitochondrial ATPase inhibitor, and vanadate, a plasma membrane ATPase inhibitor (*see* **Note 2**).

3. Evaluation of membrane functionality and integrity by demonstrating the generation of a transtonoplast pH gradient by an active V-H^+-ATPase and/or an active V-H^+-pyrophosphatase (PPase) (*see* **Note 3**).

4. Assay of the uptake of alkaloids by tonoplast vesicles followed by recovery of tonoplast vesicles by ultracentrifugation, alkaloid extraction, and analysis by HPLC-DAD. This assay is performed in the presence of an ABC transporter inhibitor and in the presence of agents disturbing the transmembrane pH gradient to establish the nature of the transporter involved.

5. Spectrofluorometric, indirect assay of transport using the pH-sensitive probe ACMA, since a dependence on the transmembrane pH gradient was observed.

2 Materials

2.1 Plant Materials

1. Plants of *C. roseus* (L.) G. Don cv. Little Bright Eye are grown at 25 °C, under a 16 h photoperiod, using light with a maximum intensity of 70 µmol $m^{-2} s^{-1}$. Seeds are acquired from AustraHort (Australia) and voucher specimens are deposited at the Herbarium of the Department of Biology of the Faculty of Sciences of the University of Porto (PO 61912). Plants used for isolation of tonoplast vesicles are 4–8-month-old, flowering plants.

2. Plants of *Nicotiana tabacum* cv. Petite Havana SR1 are grown at 21 °C, under a 12 h light photoperiod, at 35 µmol $m^{-2} s^{-1}$.

Plants suitable for isolation of tonoplast vesicles are 6–8-week-old, nonflowering plants.

3. Cell suspension cultures of *Nicotiana tabacum* (L.) cv. Bright Yellow 2 (BY-2; Riken BRC) are grown at 26.5 °C, in the dark, on an orbital shaker, at 125 rpm, in Linsmaier and Skoog medium composed of MS macronutrients and micronutrients, 200 mg L^{-1} KH$_2$PO$_4$, 100 mg L^{-1} myo-inositol, 1 mg L^{-1} thiamine·HCl, 0.2 mg L^{-1} 2,4-dichlorophenoxyacetic acid, and 3 % sucrose, pH 5.8. Cells are subcultured every 7 days by adding 2.5 mL of the full grown culture to 47.5 mL of fresh medium in 250 mL flasks.

2.2 Isolation of Tonoplast Vesicles

The buffers and solutions should be prepared no more than a few days prior to isolation of the vesicles.

1. Extraction buffer: 250 mM sucrose, 100 mM KCl, 2 mM EDTA, 3 mM MgCl$_2$, 70 mM Tris–HCl, pH 8. Filter-sterilize and store at 4 °C. Immediately before use add 1 mM phenylmethylsulfonyl fluoride (PMSF; from a 100× stock solution prepared in 100 % ethanol, stored at –20 °C), 0.1 % (w/v) bovine serum albumin (BSA), and 2 mM dl-dithiothreitol (DTT). BSA and DTT may be directly added to the buffer or prepared separately in stock solutions, aliquoted and stored at –20 °C. Aliquots should not be thawed more than once.

2. Resuspension buffer: 15 % (v/v) glycerol, 1 mM EDTA, 20 mM Tris–HCl, pH 7.5. Filter-sterilize and store at 4 °C. Immediately before use add 1 mM PMSF, and 1 mM DTT.

3. Solutions for sucrose step gradient (made fresh): 32 % or 46 % (w/v) sucrose, 1 mM EDTA, 20 mM Tris–HCl, pH 7.5. Immediately before use add 1 mM PMSF, and 1 mM DTT.

4. Reagents for protein quantification with Lowry method [9].

5. Liquid nitrogen.

6. Ultra-Turrax T25 apparatus (IKA, Janke & Kunkel, Germany).

7. Glass potter homogenizer (40 mL vessel, tight, WHEATON, USA).

8. Cheesecloth.

9. High capacity centrifuge (10,000×g).

10. Ultracentrifuge (80,000–100,000×g) with swing-out rotor.

11. Lyophilizer.

2.3 Western Blot

1. 50 mM Tris-HCl, pH 7.5.

2. Reagents for protein quantification with Bradford method [10], 10 % acrylamide gels, nitrocellulose membranes.

3. Primary antibodies: ER marker—rabbit anti-serum raised against calreticulin (a gift from J. Denecke, University of

Leeds), at a 1:10,000 dilution, or rabbit anti-serum raised against STM1 (#AS 07266, AGRISERA, Germany) at a 1:1000 dilution; chloroplast marker—rabbit anti-serum raised against the chloroplast inner envelope TIC 40 protein (#AS 10709-10, AGRISERA, Germany), at a 1:2500 dilution; tonoplast markers—rabbit anti-sera raised against a V-H$^+$-ATPase (#AS 07213, AGRISERA, Germany) and against a V-H$^+$-PPase [11] (may be requested to Prof. M. Maeshima—maeshima@agr.nagoya-u.ac.jp), both at a 1:2000 dilution.

4. Secondary antibody: peroxidase conjugated goat anti-rabbit (Santa Cruz Biotechnology, Inc) at 1:7500 dilution.

5. Chemiluminescent substrate ECL™ (GE Healthcare, Lifesciences), water bath.

2.4 V-H$^+$-ATPase Hydrolytic Activity

1. Reagents: ATP, sodium azide, sodium orthovanadate (prepared as in Subheading 2.5), Triton X-100, KCl, sodium molybdate, $MgSO_4$ in 30 mM Tris-MES pH 8, TCA, perchloric acid, NaH_2PO_4.

2. Reaction solution: 3 mM ATP, 0.02 % Triton X-100, 50 mM KCl, 1 mM sodium molybdate, and 6 mM $MgSO_4$ in 30 mM Tris-MES, pH 8.

3. Stop solution: 10 % TCA with 4 % perchloric acid.

4. Ames solution [12]: 1 volume 10 % ascorbic acid, 6 volumes of 4.2 g ammonium molybdate, and 28.6 mL H_2SO_4 in 1 L H_2O.

5. 10 mM NaH_2PO_4 with 0.02 % Triton X-100.

6. Incubator with agitation.

7. Table top centrifuge ($2400 \times g$).

8. Spectrophotometer (O.D. 820 nm).

9. Glass tubes washed with phosphate-free solutions (Milli-Q water and ethanol 100 %).

2.5 Activity of Proton Pumps

1. Reaction buffer for vesicles isolated from *C. roseus* and *N. tabacum* leaves: 30 mM KCl, 50 mM NaCl, 20 mM Hepes, pH 7.2. Filter-sterilize and store at 4 °C.

2. Reaction buffer for vesicles isolated from BY-2 cell cultures: 100 mM KCl, 10 mM MOPS-Tris, pH 7.2. Filter-sterilize and store at 4 °C.

3. 200 mM adenosine 5′-triphosphate dipotassium salt (ATP, A8937 Sigma-Aldrich) prepared in 200 mM Bis-Tris Propane—this solution will have directly a final pH of 7.2. Prepare aliquots and store at −20 °C. The aliquots should not be thawed more than once.

4. 20 mM potassium pyrophosphate (PPi, 32243-1 Sigma-Aldrich) stock solution prepared in H_2O. Store at −20 °C.

5. 2 mM ACMA (A5806 Sigma-Aldrich) stock solution prepared in 100 % ethanol. Store at −20 °C.

6. 500 mM $MgCl_2 \cdot 6H_2O$ stock solution (Merck) prepared in H_2O.

7. Stock solutions of inhibitors and uncoupling agents: 150 mM NH_4Cl (Merck), 50 mM KNO_3 (221295, Sigma-Aldrich), 50 mM sodium azide (S8032 Sigma-Aldrich), 10 mM concanamycin A (27689 Sigma-Aldrich), 50 mM sodium orthovanadate (S6508 Sigma-Aldrich).

8. Spectrofluorometer with excitation and emission wavelengths set at 415 and 485 nm respectively.

9. 2 mL UV spectrofluorometer cuvette with small magnetic stirrer.

2.6 Uptake Assay

1. Reaction buffers as in Subheading 2.5.

2. Stock solutions: 1 M DTT, 500 mM sodium creatine phosphate dibasic tetrahydrate (27920 Sigma-Aldrich), 50 mM CCCP (carbonyl cyanide *m*-chlorophenyl hydrazine; Sigma-Aldrich) in methanol; ATP, $MgCl_2$, inhibitors, and uncoupling agents stock solutions (Subheading 2.5); specialized metabolites (Subheading 2.7).

3. Creatine Kinase MM Fraction from human heart (C9983-100UG Sigma-Aldrich).

4. Ultracentrifuge ($100,000 \times g$).

5. Gaseous N_2 for sample drying.

2.7 Proton Antiport Assay

1. Reaction buffers, ATP, PPi, ACMA, and $MgCl_2$ stock solutions as in Subheading 2.5.

2. Specialized metabolites stock solutions prepared in methanol (sonicate if necessary): 100 mM vindoline (Pierre Fabre), 9 mM catharanthine (Pierre Fabre), 33 mM papaverine (P-3510 Sigma-Aldrich), 25 mM ajmaline (A-7252 Sigma-Aldrich), 100 mM atropine (A-0132 Sigma-Aldrich), 6.5 mM nicotine (N3876 Sigma-Aldrich).

3. Spectrofluorometer with excitation and emission wavelengths set at 415 and 485 nm, respectively.

4. 2 mL UV spectrofluorometer cuvette with small magnetic stirrer.

3 Methods

3.1 Isolation of Tonoplast Vesicles

Tonoplast vesicles are isolated according to [13] and [14] with minor modifications. All the procedures are carried out at 4 °C and all materials are precooled.

1. For plants (*see* **Note 4**), homogenize ~5 g of young fully developed leaves without the midrib in 100 mL of extraction buffer,

using the Ultra-Turrax T25 apparatus, at 13,000 rpm, for pulses of 20 s, over a period of about 5 min, at 4 °C (precool the beaker at 4 °C for 30 min). During the homogenization keep the beaker surrounded by ice. Repeat two times to obtain an extract from 15 g of leaves. For BY-2 cells, centrifuge 5×50 mL of 5 days old cultures at 1500×g, for 3 min, at 4 °C, wash the cells twice with cold Milli-Q water, resuspend in 300 mL of extraction buffer and homogenize with the Ultra-Turrax as above.

2. Filter the homogenate through four layers of cheesecloth, centrifuge at 10,000×g for 10 min, at 4 °C, and collect the supernatant.

3. Centrifuge the supernatant at 100,000×g for 1 h, at 4 °C, and release the pellets into a total volume of 8 mL of resuspension buffer.

4. Transfer the pellets in resuspension buffer to a glass potter homogenizer and perform ~60 up-and down cycles avoiding the formation of air bubbles, until the pellets are completely resuspended.

5. Overlay the homogenate on a 32 %/46 % sucrose step gradient with the relative proportions of 2.5:5:3, and centrifuge at 80,000×g for 3 h, at 4 °C, in a swing-out rotor, using an average deceleration.

6. Remove the upper layer (0 %) and then collect the 0 %/32 % interface containing the tonoplast vesicles and transfer to another ultracentrifuge tube.

7. Dilute in 4 volumes of resuspension buffer and centrifuge at 100,000×g for 45 min, at 4 °C.

8. Resuspend the membrane pellet in 0.3–1 mL of resuspension buffer (without PMSF or DTT), aliquot, freeze in liquid nitrogen, and store at −80 °C until use. After thawing, do not freeze again, as activity will be lost.

3.2 Characterization of Vesicles Purity by Western Blot

1. Lyophilize the vesicle samples, reconstitute in 50 mM Tris-HCl pH 7.5, and determine protein concentration using the Bradford method [10].

2. Mix protein samples with SDS-PAGE loading buffer, heat at 70 °C for 10 min, perform SDS-PAGE in 10 % acrylamide gels and transfer to a nitrocellulose membrane. Load 10 mg of protein for calreticulin and TIC 40 immuno-detection, and 1 mg of protein for V-H$^+$-ATPase and V-H$^+$-PPase immuno-detection.

3. Perform the Western blot and detection using standard protocols (*see* **Note 5**).

3.3 Characterization of Vesicles Purity by Detection of V-H⁺-ATPase Hydrolytic Activity (See Note 2)

The rate of ATP hydrolysis is determined by measuring the release of inorganic phosphate, according to [15], with some modifications.

1. Thaw a tonoplast vesicle sample and determine protein concentration with the method of Lowry [9].

2. Mix a maximum volume of 15 μL of vesicles corresponding to 15 μg of protein with 300 μL of reaction solution, and incubate for 30 min at 37 °C with slow agitation. Perform a reaction in the presence of 0.5 mM azide + 0.1 mM vanadate. In parallel, perform reactions using 0.2, 0.4, 0.6, 0.8, 1, and 1.2 mM NaH_2PO_4 in 0.1 % Triton X-100 as standards.

3. Stop the reaction by adding 500 μL of cold stop solution; keep samples on ice for 2 min and centrifuge at $2400 \times g$ for 3 min.

4. Mix the supernatant with 1.3 mL of Ames solution and incubate for 15 min at room temperature in the dark.

5. Read absorbance at 820 nm using a blank control performed without protein.

6. Build a calibration curve in nmol PO_4^{3-} min^{-1} and express the estimated ATPase hydrolytic activity in nmol PO_4^{3-} min^{-1} mg^{-1} protein.

7. Express the V-H⁺-ATPase activity as the percentage of ATP hydrolytic activity detected in the presence of inhibitors compared with the activity observed in their absence (*see* **Note 2**).

3.4 Vesicle Activity Assay

The generation of a transtonoplast pH gradient by the V-H⁺-ATPase and/or the V-H⁺- PPase is tested using the fluorescent probe ACMA (*see* **Note 1**) according to [13], with minor modifications (*see* **Note 3**).

1. Set the spectrofluorometer to 25 °C.

2. Add to a 2 mL UV spectrofluorometer cuvette with a small magnetic stirrer, 1 mL of reaction buffer, 10 μL of 500 mM $MgCl_2$, and a volume of tonoplast vesicles corresponding to 30–50 μg of protein for *C. roseus* and tobacco leaves, and to 5–10 μg of protein for BY-2 cells.

3. Set the spectrofluorometer to medium stirring and allow to equilibrate for 1 min.

4. Add 1 μL of 2 mM ACMA and let the fluorescence stabilize.

5. Start the reaction by adding ATP to a final concentration of 0.5–4.5 mM ATP or PPi to a final concentration of 0.0125–0.1 mM in order to activate proton pumping by the V-H⁺-ATPase or the V-H⁺-PPase, and register fluorescence with a 1 s resolution (*see* **Note 6**). If these pumps are active and the membranes are intact, a transmembrane pH gradient is generated resulting in fluorescence quenching of ACMA (Fig. 1).

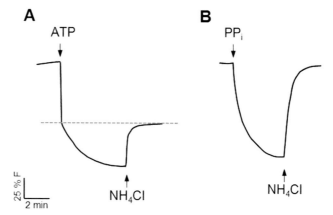

Fig. 1 Effect of the addition of ATP (**a**) and PPi (**b**) to the assay cuvette containing reaction buffer, 5 mM $MgCl_2$, 2 μM ACMA and tonoplast vesicles. 1.5 mM NH_4Cl was added as control of the dissipation of the transmembrane H^+ gradient. In (**a**), the immediate vertical quenching of ACMA upon addition of ATP is due to a direct effect on ACMA; the *dashed line* marks the beginning of the quenching due to the generation of a transmembrane pH gradient, as indicated by the dissipation of the gradient by NH_4Cl

6. Add 10 μL of 150 mM NH_4Cl. This weak base will dissipate the transmembrane pH gradient and induce a strong recovery of the ACMA fluorescence, if the previous quenching was indeed due to a transmembrane pH gradient.

7. If desired, the initial rates of fluorescence quenching for different concentration of ATP or PPi may be used to estimate the kinetic parameters of the proton pumps [4]. The rates are expressed as Δ fluorescence in % min^{-1} $μg^{-1}$ protein.

8. To further confirm the nature of the membrane vesicles, the following inhibitors may be used, added after ACMA and 3–4 min before the addition of ATP: 100 μM vanadate inhibits the P-H^+-ATPase and should not inhibit quenching if tonoplast vesicles are pure; 100 μM KNO_3 and 0.1 μM concanamycin are inhibitors of the V-H^+-ATPase and should inhibit completely the ATP induced quenching if tonoplast vesicles are pure.

3.5 Uptake Assay Using HPLC Detection

To assay the uptake of alkaloids by vesicles, they are incubated in the presence of an ATP regenerator system [16] that sustains transport through long periods of time. After incubation during different periods of time, the vesicles are collected by ultracentrifugation and alkaloids are extracted and analyzed by HPLC-DAD [4].

1. Perform a control reaction for each time point in the absence of ATP, inhibitors, and creatine phosphate and kinase. Perform reactions in the presence of different inhibitors: add the ABC transporter inhibitor vanadate to a final concentration of 1

mM, 3 min prior to ATP; add the H^+ gradient dissipator NH_4Cl to a final concentration of 1.5 mM and the protonophore CCCP to a final concentration of 50 mM, each 1 min after the addition of ATP, and 1–3 min before substrate addition (*see* **Note 7**).

2. Weigh 3 mL ultracentrifuge tubes to select pairs of tubes exactly with the same weight. You have to be very rigorous with all volume measurements, and add reaction buffer to the controls in exactly the same volumes of the specific reagents you add to other tubes. In the end, you must have exactly the same volume in all the tubes to be sure they are precisely equilibrated in weight in order to be able to proceed immediately for ultracentrifugation.

3. Add 1.2 mL of reaction buffer, 12 μL of 500 mM $MgCl_2$, and tonoplast vesicles corresponding to 100 μg of protein to the ultracentrifuge tubes on ice (*see* **Note 8**). Mix well by inverting tubes gently.

4. Add 1.2 μL of 1 M DTT, 24 μL of 500 mM creatine phosphate, and 5 μL of creatine kinase (12 μg) and mix as above.

5. Initiate energization of the tonoplast vesicles by adding 18 μL of 200 mM ATP (final 3 mM ATP) and allow energization to proceed for 3 min at 25 °C before adding 12 μL of 100 mM vindoline. Time zero tubes are always kept in ice and are centrifuged immediately after the addition of the substrate as described below in **step 7**.

6. Allow the reaction to proceed for 0, 5, 10, and 15 min at 25 °C with gentle orbital shaking.

7. To stop the reaction, transfer the tubes to ice and centrifuge immediately at $100,000 \times g$ for 30 min at 4 °C.

8. Discard the supernatant, wash the pellet superficially with ice-cold reaction buffer and dry with gaseous nitrogen. Tubes may be stored at –20 °C at this stage.

9. Resuspended each pellet in 200 μL of HPLC-grade methanol.

10. Filter and inject 20 μL in the HPLC-DAD.

11. Perform HPLC-DAD analyses on an RP-C18 column (125×4 mm) using a methanol (B) versus water (A) system containing 0.1 % (v/v) triethylamine, at a flow rate of 1 mL min^{-1}, with the following multiple step gradient : 55 % B at 0 min, 65 % B at 4 min, 67 % B at 7 min, 68.5 % B at 8 min, 70 % B at 9 min, 71.5 % B at 14 min, 90 % B at 18 min, and 55 % B at 23 min [17] (*see* **Note 9**).

12. Calculate the nmol of alkaloid present in each tube and divide by the amount of protein to express the result in nmol mg^{-1} protein. Subtract the values obtained for each control to the

transport assays, at the respective time points, to estimate the uptake of the alkaloids by the vesicles.

3.6 Proton Antiport Assay (See Note 10)

1. Start the assay by performing points 1–5 of the vesicle activity assay described in Subheading 3.4, adding ATP to obtain a final concentration of 1 mM and 4.5 mM respectively for *C. roseus* and tobacco leaf tonoplast vesicles, and adding PPi to a final concentration of 0.075 mM for tonoplast vesicles of BY-2 cells.

2. Allow fluorescence to stabilize after the ATP/PPi-induced quenching and add the substrate without interrupting fluorescence recording (*see* **Note 11**). *See* Fig. 2 for the results obtained with different substrates.

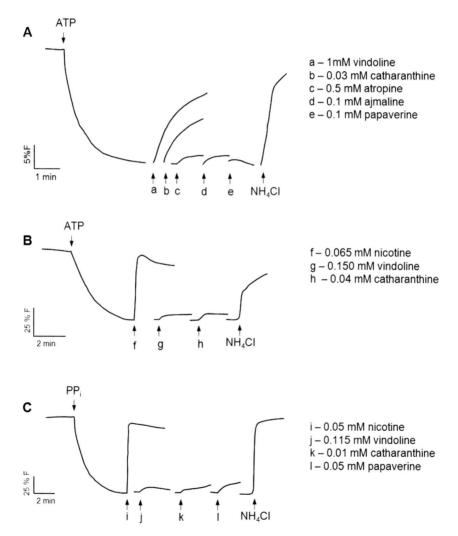

Fig. 2 Effect of the addition of specialized metabolites to a preestablished H⁺ gradient measured by the fluorescence quenching of the pH sensitive probe ACMA. Tonoplast vesicles isolated from *C. roseus* leaves (**a**), *N. tabacum* leaves (**b**) and BY-2 cells (**c**) energized with 1 mM ATP, 4.5 mM ATP, and 0.075 mM PPi, respectively, in the presence of 5 mM $MgCl_2$. 1.5 mM NH_4Cl was added as control of the dissipation of the H⁺ gradient

3. Uptake is inferred by measuring the fluorescence recovery of the ACMA probe during the first 15 s after the addition of the substrate, and rates are expressed as Δ fluorescence in % min^{-1} μg^{-1} protein.

4 Notes

1. Fluorescent amines such as ACMA are permeable through membranes in their neutral form and accumulate inside acidic compartments due to protonation. This equilibrium redistribution upon the generation of a transmembrane pH gradient (acidic inside) is associated with a quenching of their fluorescence. According to [18] this results from the fact that the only fluorescent form of ACMA is the probe in the aqueous external compartment.

2. It is postulated that the remnant ATP hydrolytic activity observed upon simultaneous inhibition with the plasma membrane H^{+}-ATPase (P-H^{+}-ATPase) inhibitor vanadate and the mitochondria H^{+}-ATPase (F-H^{+}-ATPase) inhibitor azide represents the V-H^{+}-ATPase hydrolytic activity (in the presence of Triton X100, ABC transporters lose their ATPase activity and do not interfere in this measurement). Therefore, if a ~100 % activity is observed in the presence of the two inhibitors, this indicates the absence of contamination with plasma membrane and with mitochondrial membranes, confirming a high purity of the tonoplast vesicles.

3. The generation of a transmembrane pH gradient will simultaneously demonstrate that membrane proteins are active and that the membrane is impermeable to protons, conserving its main in vivo properties.

4. Plants of *C. roseus* were 4–8 months, and all leaves from the apex until the fifth pair were used, removing the midrib from all leaves except from the two very small first pairs. Homogenization was performed in cycles of 5 g to avoid leaving leaf pieces waiting for a long time. All leaves used were still far away from being senescent (rich in proteases). There was a seasonal effect in the activity of isolated membranes, although plants were growing in a growth chamber, with maximum activity recovered in March–June (early spring to summer) and September–October (autumn). Outside these periods it was very difficult to detect transport activity. For tobacco plants the third and fourth leaves were used.

5. Increase the SDS concentration in the transfer buffer from 0.03 to 0.1 % to increase transfer efficiency and detection of membrane proteins.

6. The addition of ATP produces a direct and immediate quenching of ACMA fluorescence independent of any transport, which should be unvalued. The quenching resulting from proton pumping appears later with a slower slope (Fig. 1a). The part of the quenching corresponding to proton pumping by the V-H$^+$-ATPase is confirmed by the dissipation of the transmembrane pH gradient by addition of NH$_4$Cl (Fig. 1a). This artifact due to the addition of ATP does not happen upon addition of PPi (Fig. 1b), and is excluded from final figures (Fig. 2).

7. This assay is performed in the presence of an ABC transporter inhibitor and in the presence of agents disturbing transmembrane pH gradients to establish the nature of the transporter involved. If transport is mediated by an ABC transporter it will be inhibited by vanadate and insensitive to NH$_4$Cl and CCCP since it will not depend on a pH gradient. On the contrary, if transport is mediated by a MATE transporter, being likely a H$^+$ antiport, it will be insensitive to vanadate and inhibited by NH$_4$Cl and CCCP since both dissipate the transmembrane pH gradient—the first due to permeability of the weak conjugated base NH$_3$ that reacts with H$^+$ inside the vesicles, the second because it functions as a protonophore making the membrane permeable to protons.

8. To be sure the vesicles will remain active during the reaction, you should previously assay their energization by 3 mM ATP and the ATP regeneration system, in the presence of ACMA, for the maximum period of time you will be using. Add DTT, creatine phosphate and creatine phosphatase before the ACMA and the 3 mM ATP after the ACMA.

9. For the detection of transport of other specialized metabolites it should not be difficult to find or establish an HPLC method for their quantification.

10. Since alkaloid uptake by tonoplast vesicles indicated a dependence on a transmembrane pH gradient, the alkaloid transport system was further characterized using the fluorescent probe ACMA (*see* **Note 1**). The ACMA assay is much faster than the uptake assay with HPLC detection, and uses a significantly lower amount of vesicles. Therefore, it is much more convenient to investigate specificity, kinetics, etc. In fact, it can be performed before the uptake assay, as soon as the functionality of vesicles is checked with ACMA (Subheading 3.3). However, it should be taken in account that this assay only provides indirect evidence and transport should always be directly measured by an uptake assay.

11. Depending on the apparatus, this may cause variable recording artifacts that may be erased afterwards, but you should not stop recording, otherwise you may loose the initial rate of fluorescence recovery.

Acknowledgements

This work was supported by: (1) Fundo Europeu de Desenvolvimento Regional funds through the Operational Competitiveness Programme COMPETE and by National Funds through Fundação para a Ciência e a Tecnologia (FCT) under the projects FCOMP-01-0124-FEDER-037277 (PEst-C/SAU/LA0002/2013) and FCOMP-01-0124-FEDER-019664 (PTDC/BIA-BCM/119718/2010); (2) by the FCT scholarships co-supported by FCT and POPH-QREN (European Social Fund), SFRH/BD/41907/2007 (IC) and SFRH/BD/74257/2010 (HN); (3) by a Postdoctoral fellowship financed by national funds through FCT under the project Incentivo/SAU/LA0002/2014 (VM); (4) by a Scientific Mecenate Grant from Grupo Jerónimo Martins.

References

1. Shoji T (2014) ATP-binding cassette and multidrug and toxic compound extrusion transporters in plants: a common theme among diverse detoxification mechanisms. Int Rev Cell Mol Biol 309:303–346

2. Shitan N, Yazaki K (2013) New insights into the transport mechanisms in plant vacuoles. Int Rev Cell Mol Biol 305:383–433

3. Conde A, Regalado A, Rodrigues D, Costa JM, Blumwald E, Chaves MM, Gerós H (2015) Polyols in grape berry: transport and metabolic adjustments as a physiological strategy for water-deficit stress tolerance in grapevine. J Exp Bot 66:889–906

4. Carqueijeiro I, Noronha H, Duarte P, Gerós H, Sottomayor M (2013) Vacuolar transport of the medicinal alkaloids from *Catharanthus roseus* is mediated by a proton driven antiport. Plant Physiol 162:1486–1496

5. Martins V, Hanana M, Blumwald E, Gerós H (2012) Copper transport and compartmentation in grape cells. Plant Cell Physiol 53:1866–1880

6. Johansson F, Olbe M, Sommarin M, Larsson C (1995) Brij 58, a polyoxyethylene acyl ether, creates membrane vesicles of uniform sidedness. A new tool to obtain inside-out (cytoplasmic side-out) plasma membrane vesicles. Plant J 7:165–173

7. Conde A, Diallinas G, Chaumont F, Chaves M, Gerós H (2010) Transporters, channels or simple diffusion? Dogmas, atypical roles and complexity in transport systems. Int J Biochem Cell Biol 42:857–868

8. Conde C, Agasse A, Glissant D, Tavares R, Gerós H, Delrot S (2006) Pathways of glucose regulation of monosaccharide transport in grape cells. Plant Physiol 141:1563–1577

9. Lowry OH, Rosebrough NJ, Farr AL, Randall RJ (1951) Protein measurement with the Folin phenol reagent. J Biol Chem 193:265–275

10. Bradford MM (1976) A rapid and sensitive method for the quantitation of microgram quantities of protein utilizing the principle of protein-dye binding. Anal Biochem 72:248–254

11. Maeshima M, Yoshida S (1989) Purification and properties of vacuolar membrane proton-translocating inorganic pyrophosphatase from mung bean. J Biol Chem 264:20068–20073

12. Ames BN (1966) Assay of inorganic phosphate, total phosphate and phosphatases. Methods Enzymol 8:115–118

13. Façanha AR, de Meis L (1998) Reversibility of H^+-ATPase and H^+-pyrophosphatase in tonoplast vesicles from maize coleoptiles and seeds. Plant Physiol 116:1487–1495

14. Queirós F, Fontes N, Silva P, Almeida DPF, Maeshima M, Gerós H, Fidalgo F (2009) Activity of tonoplast proton pumps and Na^+/H^+ exchange in potato cell cultures is modulated by salt. J Exp Bot 60:1363–1374

15. Vera-Estrella R, Barkla BJ, Higgins VJ, Blumwald E (1994) Plant defense response to fungal pathogens: activation of host-plasma membrane H^+-ATPase by elicitor-induced enzyme dephosphorylation. Plant Physiol 104:209–215

16. Marinova K, Pourcel L, Weder B, Schwarz M, Barron D, Routaboul JM, Debeaujon I, Klein M (2007) The Arabidopsis MATE transporter TT12 acts as a vacuolar flavonoid/H⁺-antiporter active in proanthocyanidin- accumulating cells of the seed coat. Plant Cell 19:2023–2038

17. Sottomayor M, dePinto MC, Salema R, DiCosmo F, Pedreno MA, Barcelo AR (1996) The vacuolar localization of a basic peroxidase isoenzyme responsible for the synthesis of alpha-3',4'-anhydrovinblastine in *Catharanthus roseus* (L) G. Don leaves. Plant Cell Environ 19:761–767

18. Casadio R (1991) Measurements of transmembrane pH differences of low extents in bacterial chromatophores. A study with the fluorescent probe 9-amino, 6-chloro, 2-methoxyacridine. Eur Biophys J 19:189–201

Chapter 13

Protoplast Transformation as a Plant-Transferable Transient Expression System

Patrícia Duarte, Diana Ribeiro, Inês Carqueijeiro, Sara Bettencourt, and Mariana Sottomayor

Abstract

The direct uptake of DNA by naked plant cells (protoplasts) provides an expression system of exception for the quickly growing research in non-model plants, fuelled by the power of next-generation sequencing to identify novel candidate genes. Here, we describe a simple and effective method for isolation and transformation of protoplasts, and illustrate its application to several plant materials.

Key words Plant-transferable methodology, PEG-mediated transformation, Protoplasts, Transient expression

1 Introduction

Protoplasts were first isolated in 1960 by Cocking and, since then, they revealed to be a unique cell system for the study of many cellular processes, as well as a successful biotechnological tool for somatic hybridization breeding programs [1–3]. Notably, transient expression in protoplasts has been a very useful tool for investigation of signal transduction pathways, subcellular sorting, transcriptional regulation, protein–protein interactions, activity characterization of recombinant proteins, etc. [1, 4, 5]. In the near future, this technique should further grow in importance due to the advent of next-generation sequencing, which allows the collection of massive amounts of gene sequence data about any species of interest. As an outcome of this, the study of gene function can now be expanded from a limited number of model species to any conceivable target plant, provided customized molecular tools are available. Hence, methodologies such as transient expression in protoplasts, that are easily plant-transferable, will no doubt become key tools for much emerging research in non-model plants.

Arthur Germano Fett-Neto (ed.), *Biotechnology of Plant Secondary Metabolism: Methods and Protocols*, Methods in Molecular Biology, vol. 1405, DOI 10.1007/978-1-4939-3393-8_13, © Springer Science+Business Media New York 2016

Transient expression in protoplasts is a highly efficient but simple technique that requires no special equipment, it is species independent, and it generates results within hours or days. The simplest method to enable successful uptake of DNA by protoplasts is the use of polyethylene glycol (PEG) [6], being widely used and easily applied to different species [7–9]. On the other hand, the isolation of protoplasts has to be customized for each species/organ/tissue, having in mind that protoplasts largely retain the cell identity and differentiated state of their source cells [10] and is, therefore, an important feature to take in account for the design of gene function screenings. Methods for isolation of protoplasts from many different plants/organs/tissues are available in the literature, and it should not be difficult to develop a protoplast isolation method for most plant species/organs/tissues [2]. Finally, it is also noteworthy that transgene expression in protoplasts may ultimately be used for the establishment of stable transformants, through selective long-term protoplast culture, enabling to generate stably transformed cell lines or transgenic plants, if a regeneration protocol is available [2, 3].

Here, we describe in detail: (1) a simple and effective method for isolation of protoplasts obtained from cell suspensions of *Vitis vinifera*, leaves of tobacco, and leaves of the alkaloid producing plant *Catharanthus roseus*, in which the single species customized part of the procedure was the enzyme composition of the cell wall digestion medium, and (2) a PEG-mediated protocol for protoplast transformation, based on [6], yielding high efficiency expression rates for the three biological materials. We also include a description of staining with calcofluor-white (CFW) to test for the presence of cell wall during protoplast isolation, and with fluorescein diacetate (FDA) to evaluate protoplast viability.

2 Materials

2.1 Plant Materials

1. Plants of *C. roseus* (L.) G. Don cv.F Little Bright Eye were grown at 25 °C, under a 16 h photoperiod, using light with a maximum intensity of 70 μmol/m²/s. Seeds were acquired from AustraHort (Australia) and voucher specimens are deposited at the Herbarium of the Department of Biology of the Faculty of Sciences of the University of Porto (PO 61912). Plants used for protoplast isolation and transformation were 6–8 months old.

2. Plants of *Nicotiana tabacum* cv. Petite Havana SR1 were grown at 21 °C, under a 12 h light photoperiod, at 35 μmol/m²/s. Plants suitable for protoplast isolation and transformation were 6–8 weeks old.

3. Cell suspension cultures of *V. vinifera* (L.) Cabernet Sauvignon Berry [11] were grown at 25 °C, in the dark, on an orbital

shaker at 125 rpm, in CBS medium composed of MS macro and micronutrients, 100 mg/L myo-inositol, 1 mg/L nicotinic acid, 1 mg/L calcium D pantothenate, 0.01 mg/L D-biotin, 1 mg/L pyridoxine HCl, 1 mg/L thiamine, 0.46 mg/L 1-naphthaleneacetic acid, 0.12 mg/L 6-benzylamino-purine, 250 mg/L casein, and 2 % sucrose, pH 5.8. Cells were subcultured every 7 days adding 10 mL of the full grown culture to 40 mL of fresh medium in 250 mL flasks.

2.2 Isolation of Protoplasts

1. MM buffer (see **Notes 1** and **2**): 0.4 M mannitol in 20 mM MES, adjust pH to 5.7 with 1 M KOH. Autoclave and store at 4 °C.

2. Digestion medium (see **Note 3**): 2 % (w/v) cellulase Onozuka R-10 (Duchefa), 0.3 % (w/v) macerozyme R-10 (Serva), and 0.1 % (v/v) pectinase (Sigma) dissolved in MM buffer. Stir very gently at least for 30 min, in the dark. Prepare immediately prior to tissue harvest. For tobacco, digestion medium is composed of 0.4 % (w/v) cellulase Onozuka R-10 and 0.2 % (w/v) macerozyme R-10.

3. MMg buffer: 0.4 M mannitol and 15 mM $MgCl_2$ in 4 mM MES, adjust pH to 5.7 with 1 M KOH. Autoclave and store at 4 °C.

4. Desiccator.

5. Vacuum pump.

6. Orbital shaker.

7. Hemocytometer.

8. Optical microscope.

9. 100 μm nylon mesh; disposable plastic Pasteur pipettes; carbon steel surgical blades.

10. Plastic Petri dishes.

11. 15 mL plastic centrifuge tubes.

2.3 Staining of Protoplasts with Calcofluor-White (CFW) and Fluorescein Diacetate (FDA)

1. 1 % w/v CFW in water (see **Note 4**).

2. 10 μg/μL FDA in acetone (see **Note 5**).

3. Fluorescence microscope with filters for (1) excitation wavelengths of 387 ± 11 nm and emission wavelengths of 457 ± 22 nm, (2) excitation wavelengths of 494 ± 20 nm and emission wavelengths of 530 ± 20 nm.

4. Slides and coverslips.

2.4 Transformation of Protoplasts Mediated by PEG

1. Ultrapure plasmid DNA (see **Note 6**).

2. PEG solution: 40 % (w/v) PEG (Sigma, ref. 81240), 0.2 M mannitol, and 0.1 M $CaCl_2$. Filter-sterilize and store at room temperature for up to 5 days.

3. W5 solution: 154 mM NaCl, 125 mM CaCl$_2$, and 5 mM KCl in 2 mM MES, adjust pH to 5.7 with 1 M KOH. Autoclave and store at 4 °C.

4. Round-bottom 2 mL tubes; 15 mL plastic centrifuge tubes.

---

3   Methods

3.1   Isolation of Protoplasts from C. roseus Leaves

Isolation of protoplasts was optimized taking in account some methods previously described [6, 12–14].

1. Cut 8–10 (~1.5–2 g) fully developed young leaves of healthy adult plants into ~1 mm strips, excluding the central vein (*see* **Note 7**). Soak the blade with the medium several times as you cut and change the blade every 3–4 leaves.

2. Immediately transfer the leaf strips, abaxial face down, to a Petri dish with 10 mL of digestion medium.

3. Vacuum infiltrate the medium on the open Petri dish for 15 min, applying slow disruptions of the vacuum every 30 s using the desiccator and the vacuum pump.

4. Incubate the leaf strips in the digestion medium for ca. 3 h at 25 °C, in the dark (*see* **Note 8**).

5. After this incubation, place the Petri dishes on an orbital shaker (~60 rpm) for 15 min in the dark and at room temperature (RT), to help release the protoplasts.

6. Release the protoplasts from leaf strips with the help of a sawn-off plastic Pasteur pipette, by gently pressing the leaf strips against the side wall of the Petri dish and slowly flushing the medium over them.

7. Filter the protoplasts through a 100 µm nylon mesh into a new Petri dish, maintaining the mesh in contact with the dish to guarantee a continuous flow of solution and avoid dripping. Wash the leaf strips with 2 mL of MM buffer to recover the maximum number of protoplasts.

8. Gently transfer the protoplasts to two 15 mL plastic tubes, using sawn-off plastic Pasteur pipettes, always avoiding dripping. Wash the Petri dish with 2 mL of MM buffer.

9. Centrifuge the protoplast suspension at 65 × *g* with Acc/Des 1 for 5 min at RT to pellet the protoplasts. Remove the supernatant with a pipette and discard it.

10. Wash the protoplasts of each tube three times with 5 mL of MM buffer: resuspend the protoplasts gently, by flicking the tube after addition of a small volume of medium (1 mL), followed by the addition of the remaining washing volume and centrifugation as in **step 9**.

11. The last pellets are all resuspended in a minimum volume of MM buffer and pooled together. Determine the protoplast concentration using a hemocytometer under the optical microscope and in the meantime let the protoplasts rest for 30 min at RT.

12. After this time, protoplasts settle at the bottom of the tube by gravity. Discard the supernatant and resuspend the protoplasts in the adequate volume of MMg buffer, for a final concentration of 5×10^6 cells/mL. At this point, the protoplasts are ready for transformation.

Images of *C. roseus* leaf protoplasts can be seen in Fig. 1.

3.2 Isolation of Protoplasts from N. tabacum Leaves

N. tabacum mesophyll protoplasts are isolated using the methodology employed for *C. roseus* with the following modifications:

1. Digestion medium has a specific composition (*see* Subheading 2.2, **item 2**).

2. Protoplasts are not shaken at any point.

Images of *N. tabacum* leaf protoplasts can be seen in Fig. 1.

3.3 Isolation of Protoplasts from V. vinifera Cell Suspensions

Protoplasts from *V. vinifera* suspension cells are isolated from 50 mL of a 2–3 day-old suspension culture.

1. Sediment the cells by centrifugation at $100 \times g$ for 5 min at RT and resuspend in 50 mL of the digestion medium used for *C. roseus*.

2. Transfer the cell suspension to a 140 mm diameter Petri dish and incubate for ca. 3 h at 25 °C, in the dark, on an orbital shaker at 25 rpm.

3. Hereafter the procedure is the same as described for *C. roseus*.

Images of protoplasts from *V. vinifera* suspension cells can be seen in Fig. 1.

3.4 Staining of Protoplasts with Calcofluor-White

The generation of cell wall free protoplast cells can be followed by using CFW staining (Fig. 2) (*see* **Note 4**).

1. Add 2 μL of CFW to ~5×10^5 protoplasts (100 μL of cells diluted with 400 μL of MM).

2. Incubate for 10 min in the dark, and observe under a fluorescence microscope using a filter with excitation wavelengths of 387 ± 11 nm and emission wavelengths of 457 ± 22 nm.

3.5 Staining of Protoplasts with Fluorescein Diacetate

The viability of the isolated protoplasts can be measured by using FDA (Fig. 3) (*see* **Note 5**).

1. Add 5 μL of FDA to ~5×10^5 protoplasts (100 μL of cells diluted with 400 μL of MM).

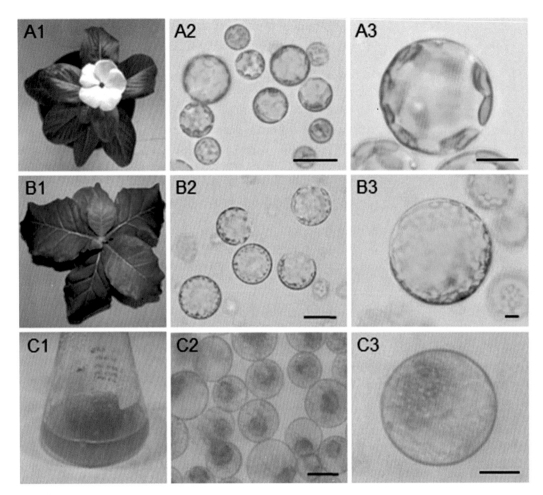

Fig. 1 Bright-field images of protoplasts from *C. roseus* leaves (**A2** and **A3**), *N. tabacum* leaves (**B2** and **B3**) and *V. vinifera* suspension cells (**C2** and **C3**) at different magnifications. (**A1**) *C. roseus* plants at the developmental stage used for protoplast isolation. (**B1**) *N. tabacum* plants at the developmental stage used for protoplast isolation. (**C1**) *V. vinifera* cell suspension culture. Bars = 30 μm (**A2**, **B2**, and **C2**), 10 μm (**A3**, **B3**, and **C3**)

2. Incubate for 10 min in the dark, and observe under a fluorescence microscope using a filter with excitation wavelengths of 494 ± 20 nm and emission wavelengths of 530 ± 20 nm.

3.6 Transformation of Protoplasts Mediated by PEG

Transformation of *C. roseus*, *N. tabacum* and *V. vinifera* protoplasts was adapted from [6].

1. Add 10–20 μg of the plasmid DNA to round-bottom 2 mL tubes (*see* **Note 6**).

2. Add 100 μL (5 × 10⁵ cells) of the protoplast suspension (using a sawn off P1000 micropipette tip) to the DNA containing tube and mix well.

3. Slowly add 110 μL of the PEG solution to the DNA–protoplast mixture, drop by drop, allowing each drop to slide through

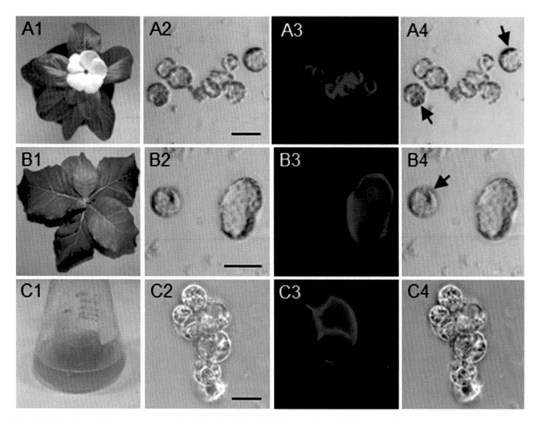

Fig. 2 Detection of the presence of cell wall with calcofluor-white (CFW) 1 h after the beginning of cell wall digestion of *C. roseus* leaf cells (**A2–A4**), *N. tabacum* leaf cells (**B2–B4**) and *V. vinifera* suspension cells (**C2–C4**). (**A1**) *C. roseus* plants at the developmental stage used for protoplast isolation. (**B1**) *N. tabacum* plants at the developmental stage used for protoplast isolation. (**C1**) *V. vinifera* cell suspension culture. (**A2**, **B2**, and **C2**) Bright-field images. (**A3**, **B3**, and **C3**) UV light images. (**A4**, **B4**, and **C4**) Merge of images 2 and 3 for each organism. Blue fluorescence under UV light corresponds to cell wall stained with CFW. *Arrows* indicate perfectly spherical cells corresponding to isolated protoplasts without cell wall. Bars = 30 μm

the side wall of the tube and gently flicking the tube after each drop, just enough to mix well (*see* **Notes 9** and **10**).

4. Leave the tubes incubating on the bench for 5–15 min.

5. After this incubation time, add 440 μL of W5 solution and mix as before.

6. Spin down the protoplasts at $65 \times g$ with Acc/Des 1 for 2 min.

7. Remove the supernatant and resuspend the pellet in 100 μL of W5 solution.

8. Transfer the samples (using sawn off P1000 micropipette tips) into 15 mL falcon tubes containing 900 μL of W5 solution.

9. Incubate in the dark at 25 °C for the desired period of time (*see* **Note 11**) with the tubes in a slight slope (*see* **Note 12**).

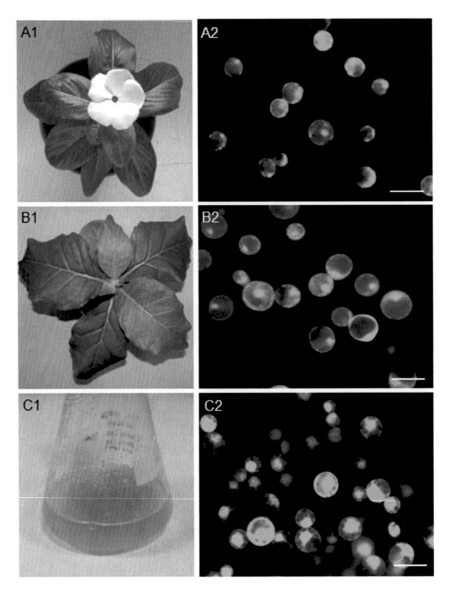

Fig. 3 Viability of isolated protoplasts upon staining with fluorescein diacetate (FDA) for *C. roseus* leaf protoplasts (**A2**), *N. tabacum* leaf protoplasts (**B2**) and *V. vinifera* suspension cell protoplasts (**C2**). (**A1**) *C. roseus* plants at the developmental stage used for protoplast isolation. (**B1**) *N. tabacum* plants at the developmental stage used for protoplast isolation. (**C1**) *V. vinifera* cell suspension culture. Green fluorescence reveals viable protoplasts. Bars = 20 μm

10. After the customized period of time, perform your observation or analysis of interest. You may harvest protoplasts by centrifugation at 65 × *g* for 2 min and freeze them until further use. You may either freeze dry or maintain frozen and resuspend on the desired buffer or solvent upon thawing. Images of protoplasts from the three species transiently expressing GFP can be seen in Fig. 4.

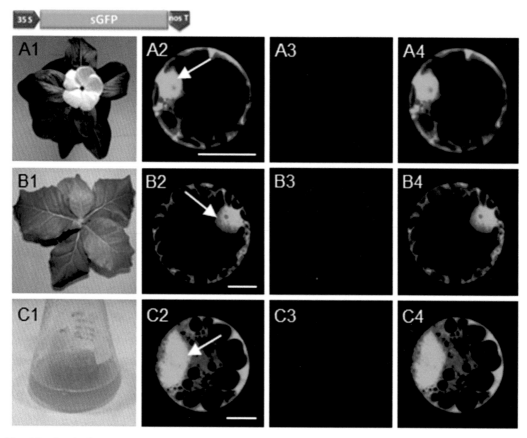

Fig. 4 Confocal microscope images of *C. roseus* leaf protoplasts (**A2–A4**), *N. tabacum* leaf protoplasts (**B2–B4**) and *V. vinifera* suspension cell protoplasts (**C2–C4**) transiently transformed with the GFP marker construct 35S::sGFP showing fluorescence in the cytosol and the nucleoplasm (*arrows*). The schematic representation of the construct is depicted *right* above the set of images. (**A1**) *C. roseus* plants at the developmental stage used for protoplast isolation. (**B1**) *N. tabacum* plants at the developmental stage used for protoplast isolation. (**C1**) *V. vinifera* cell suspension culture. Column 2 corresponds to the GFP channel, column 3 to the red channel showing chloroplast autofluorescence and column 4 depicts the merged images for each respective line. Bars = 10 μm

4 Notes

1. The osmotic pressure stabilizing agents and the presence of protecting agents such as BSA, $CaCl_2$, and KCl in the digestion buffer are important factors in protoplast isolation. The most used osmoticum is mannitol and, here, we obtained excellent results for several plant materials with the MM buffer from [13], which includes no protectant, including only mannitol and MES buffer.

2. The concentration of the osmoticum is important, with the reported rule of thumb being to use an osmoticum concentration slightly above the point of incipient plasmolysis. Many reports even include a preplasmolysis short incubation in the medium without enzymes prior to digestion as an accelerating/homogenizing step [8, 15]. However, although this was performed in [12] for *C. roseus* leaves, with the use of 0.6 M sorbitol, very good results were also obtained here using 0.4 M mannitol.

3. The key step in protoplast isolation is the enzyme mixture used for cell wall digestion. The combination of cellulase Onozuka RS and pectolyase Y23 has been claimed to be suitable for the isolation of protoplasts of a high spectrum of species and tissues, due to their high and broad activities [16]. These enzymes indeed work quite well with *C. roseus* mesophyll cells (data not shown) but they are very expensive, and good results were also obtained here using a balanced mixture of the cheaper cellulase Onozuka R10, macerozyme R10, and pectinase. However, the use of highly active enzymes allows the use of lower concentrations, what may be important for sensitive cells susceptible to enzyme toxicity. Concentrations of cellulase are usually between 0.5 and 3 %, with concentrations of macerozyme and pectinase being usually between 0.1 and 0.4 %, although they can be raised at least up to 1.5 % for recalcitrant tissues [15].

4. Calcofluor-white is a special fluorescent stain that binds strongly to structures containing cellulose, thus it is used for the staining of cell walls of both algae and higher plants.

5. The intact plasma membrane is permeable to FDA, which is accumulated inside the cells due to conversion to a membrane impermeable green fluorescent dye (fluorescein) by cytoplasmic esterases. Thus, accumulation of fluorescein in the cytoplasm is a measure of two independent processes, membrane integrity and the presence of active esterases, and represents a highly reliable indication of cell viability [11].

6. The pDNA to be used must be highly pure and highly concentrated (1.5–2 µg/µL). We recommend QIAGEN Plasmid Midi Kit.

7. It is important that the plants used are healthy and under controlled conditions to avoid any stress. We and others have observed a strong seasonal influence even for plants grown in growth chambers, with lower yields and quality obtained during autumn and winter.

8. It is important to improve conditions to obtain the shortest possible digestion incubation period, in order to avoid physiological changes and enzyme toxicity, which can render

protoplasts too sensitive to subsequent manipulations. Many protoplast isolation protocols do this incubation under weak orbital shaking, but with our protocol that is not necessary or advisable, possibly because we use a low osmotic concentration.

9. In some species, PEG-mediated gene transfer to protoplasts may be hampered by the release of nucleases from the protoplasts, resulting in the degradation of plasmid DNA. Folling et al. [17] reported that for transformation of *Lolium perenne* L. protoplasts, plasmid DNA was best protected by a combination of high pH and low temperature.

10. Do not handle more than six samples at a time. It takes time to add the PEG solution to the DNA–protoplast mixture and the incubation period must not exceed 15 min. So, normally, when you finish adding PEG solution to the sixth tube, the 15 min period is almost completed, and you should start adding W5 solution to the first tube.

11. Yoo et al. [6] refers that the protoplast incubation time should be 2–6 h for RNA analysis and 2–16 h for enzyme activity analysis, and that stimulus can be applied during the incubation according to the experimental purposes. For observation of fluorescent reporter proteins we incubate for 24–72 h depending on the fusion proteins [7]. You may have to perform a time course to evaluate the peak of your target.

12. This is recommended to decrease the suspension depth and prevent hypoxia stress.

Acknowledgements

This work was supported by: (1) Fundo Europeu de Desenvolvimento Regional funds through the Operational Competitiveness Programme COMPETE and by National Funds through Fundação para a Ciência e a Tecnologia (FCT) under the projects FCOMP-01-0124-FEDER-037277 (PEst-C/SAU/LA0002/2013), FCOMP-01-0124-FEDER-019664 (PTDC/BIA-BCM/119718/2010) and FCOMP-01-0124-FEDER-028125 (PTDC/BBB-BIO/2231/2012); (2) by the FCT scholarships co-supported by FCT and POPH-QREN (European Social Fund) SFRH/BD/41907/2007 (IC), SFRH/BD/48283/2008 (SB) and SFRH/BPD/20669/2004 (PD); (3) by a Scientific Mecenate Grant from Grupo Jerónimo Martins.

References

1. Sheen J (2001) Signal transduction in maize and Arabidopsis mesophyll protoplasts. Plant Physiol 127:1466–1475

2. Davey MR, Anthony P, Power JB, Lowe KC (2005) Plant protoplasts: status and biotechnological perspectives. Biotechnol Adv 23:131–171

3. Eeckhaut T, Lakshmanan PS, Deryckere D, Van Bockstaele E, Van Huylenbroeck J (2013) Progress in plant protoplast research. Planta 238:991–1003

4. Cheng SH, Sheen J, Gerrish C, Bolwell GP (2001) Molecular identification of phenylalanine ammonia-lyase as a substrate of a specific constitutively active Arabidopsis CDPK expressed in maize protoplasts. FEBS Lett 503:185–188

5. Martens S, Teerib T, Forkmanna G (2002) Heterologous expression of dihydroflavonol 4-reductases from various plants. FEBS Lett 531:453–458

6. Yoo SD, Cho YH, Sheen J (2007) *Arabidopsis* mesophyll protoplasts: a versatile cell system for transient gene expression analysis. Nat Protoc 2:1565–1572

7. Duarte P, Ribeiro D, Henriques G, Hilliou F, Rocha AS, Lima F, Amorim I, Sottomayor M (2011) Cloning and characterization of a candidate gene from the medicinal plant *Catharanthus roseus* through transient expression in mesophyll protoplasts. In: Brown GG (ed) Molecular cloning-selected applications in medicine and biology. Intech, Rijeka, pp 291–308

8. Zhang Y, Su J, Duan S, Ao Y, Dai J, Liu J, Wang P, Li Y, Liu B, Feng D, Wang J, Wang H (2011) A highly efficient rice green tissue protoplast system for transient gene expression and studying light/chloroplast-related processes. Plant Methods 7:30

9. Li JF, Chung HS, Niu Y, Bush J, McCormack M, Sheen J (2013) Comprehensive protein-based artificial microRNA screens for effective gene silencing in plants. Plant Cell 25:1507–1522

10. Faraco M, Di Sansebastiano GP, Spelt K, Koes RE, Quattrocchio FM (2011) One protoplast is not the other! Plant Physiol 156:474–478

11. Fontes N, Silva R, Vignault C, Lecourieux F, Geros H, Delrot S (2010) Purification and functional characterization of protoplasts and intact vacuoles from grape cells. BMC Res Notes 3:19

12. Sottomayor M, dePinto M, Salema R, DiCosmo F, Pedreno M, Barceló AR (1996) The vacuolar localization of a basic peroxidase isoenzyme responsible for the synthesis of alpha-3′,4′-anhydrovinblastine in *Catharanthus roseus* (L) G. Don leaves. Plant Cell Environ 19:761–767

13. Zheng HQ, Wang GL, Zhang L (1997) Alfalfa mosaic virus movement protein induces tubules in plant protoplasts. Mol Plant Microbe Interact 10:1010–1014

14. Ferreres F, Figueiredo R, Bettencourt S, Carqueijeiro I, Oliveira J, Gil-Izquierdo A, Pereira DM, Valentao P, Andrade PB, Duarte P, Barceló AR, Sottomayor M (2011) Identification of phenolic compounds in isolated vacuoles of the medicinal plant *Catharanthus roseus* and their interaction with vacuolar class III peroxidase: an H_2O_2 affair? J Exp Bot 62:2841–2854

15. Zhao W, Yang W, Wei C, Sun G (2011) A simple and efficient method for isolation of pineapple protoplasts. Biotechnol Biotechnol Equip 25:2464–2467

16. Nagata T, Okada K, Takebe I, Matsui C (1981) Delivery of tobacco mosaic-virus RNA into plant-protoplasts mediated by reverse-phase evaporation vesicles (liposomes). Mol Gen Genet 184:161–165

17. Folling M, Pedersen C, Olesen A (1998) Reduction of nuclease activity from *Lolium* protoplasts: effect on transformation frequency. Plant Sci 139:29–40

Chapter 14

Design, Construction, and Validation of Artificial MicroRNA Vectors Using *Agrobacterium*-Mediated Transient Expression System

Basdeo Bhagwat, Ming Chi, Dianwei Han, Haifeng Tang, Guiliang Tang, and Yu Xiang

Abstract

Artificial microRNA (amiRNA) technology utilizes microRNA (miRNA) biogenesis pathway to produce artificially selected small RNAs using miRNA gene backbone. It provides a feasible strategy for inducing loss of gene function, and has been applied in functional genomics study, improvement of crop quality and plant virus disease resistance. A big challenge in amiRNA applications is the unpredictability of silencing efficacy of the designed amiRNAs and not all constructed amiRNA candidates would be expressed effectively in plant cells. We and others found that high efficiency and specificity in RNA silencing can be achieved by designing amiRNAs with perfect or almost perfect sequence complementarity to their targets. In addition, we recently demonstrated that *Agrobacterium*-mediated transient expression system can be used to validate amiRNA constructs, which provides a simple, rapid and effective method to select highly expressible amiRNA candidates for stable genetic transformation. Here, we describe the methods for design of amiRNA candidates with perfect or almost perfect base-pairing to the target gene or gene groups, incorporation of amiRNA candidates in miR168a gene backbone by one step inverse PCR amplification, construction of plant amiRNA expression vectors, and assay of transient expression of amiRNAs in *Nicotiana benthamiana* through agro-infiltration, small RNA extraction, and amiRNA Northern blot.

Key words *Agrobacterium*-mediated transient expression system; artificial microRNA, amiRNA design and construction, miR168a, Plant amiRNA expression vector

1 Introduction

Artificial microRNA (amiRNA) technology exploits microRNA (miRNA) biogenesis pathway to produce artificially designed small RNAs using miRNA gene backbone [1–4]. It generates a single type of small RNA population all with the same nucleic acid sequence, usually 21 nucleotides (nt) in length, providing a feasible method for either silencing an individual gene or simultaneously silencing closely related gene isoforms [5]. Utilization of amiRNAs to induce loss of gene function has been achieved in

Arthur Germano Fett-Neto (ed.), *Biotechnology of Plant Secondary Metabolism: Methods and Protocols*, Methods in Molecular Biology, vol. 1405, DOI 10.1007/978-1-4939-3393-8_14, © Springer Science+Business Media New York 2016

diverse plant species, indicating broad potential application, but the technology requires improved predictability of silencing efficacy of the designed amiRNA candidates [1, 6, 7]. Multiple criteria for design of functional amiRNAs have been proposed [3, 8, 9], but a recent study found that silencing efficacy of designed amiRNAs was generally less potent than that of the endogenous miRNA even though the amiRNAs and the miRNA showed analogous target complementarities [6]. The web-based miRNA designer (WMD) can automatically design gene-specific amiRNA candidates for over 100 plant species [3], but an investigation found that only 15 % of the amiRNA candidates were optimal for inducing RNA silencing [7].

Compared with the WMD [3], we designed amiRNAs with some differences. The basic principle of our method is to design a small RNA of 21-nt that has perfect or almost perfect sequence complementarity to the gene of interest. It is based on a previous study by Tang et al. [10], which found that a target RNA could be efficiently cleaved at the same cleavage site by RNA-induced silencing complex (RISC) guided with either an endogenous miRNA that is not completely complementary to the target sequence or a synthetic small interfering RNA (siRNA) that perfectly base-paired to the same target sequence of the miRNA. In our design, if a perfect sequence complementarity is not available between an amiRNA and its target(s), a G:U wobble base-paring will be considered [11]. As to specific base requirement, a uracil (U) residue at position 1 of an amiRNA from the 5′-end is preferred, but an adenine (A) at the position 1 appeared functional at least in one case [5]. The tenth position of amiRNAs is suggested to be preferably an A residue by other designers [3, 8], but it is not required in our design. We noticed that miR165, 166, and 168, the tenth position is not an A but they performed cleavage perfectly and efficiently [5, 10]. Our program analyzes the free energies (ΔG) of the 5′-end of both strands of a candidate amiRNA duplex between the guide strand (amiRNA) and the passenger strand (amiRNA*). Low thermodynamic stability at the 5′-end of the amiRNA compared to the amiRNA* will increase the opportunity of incorporation of the amiRNA into RISC [12, 13]. Experimental data demonstrated that amiRNAs with perfect sequence complementarity to their target genes achieved high efficiency and specificity in RNA silencing [2, 5, 14, 15]. One concern was that perfect sequence complementarity between amiRNAs and their targets may trigger the generation of secondary small RNA [3], but a recent research concluded that perfect base-pairing between amiRNA and its target does not provoke secondary small RNA production [14, 15].

Another challenge in amiRNA applications was that not all of the designed amiRNA candidates would be expressible in plant cells. Different amiRNA sequences inserted in the same miR168a gene backbone could result in dramatically different production levels of mature amiRNAs [1, 5]. Variable expression of amiRNAs in transgenic plants and positive correlation between gene knockdown efficiency and the amiRNA levels were observed previously [4, 16–19]. It is usually labor intensive and time-consuming to make transgenic plants and screen for desired transgenic genome types. We recently demonstrated that *Agrobacterium*-mediated transient expression system can be used to screen and select effective amiRNA constructs for plant genetic transformation [1]. As a general trend, transformation of a miR168a-derived amiRNA construct that shows high expression level of amiRNAs in the transient expression system would increase the probability of obtaining transgenic lines with stably expressed amiRNAs. In this chapter, we present the methods that we developed to design amiRNAs, construct amiRNA vectors and detect the expression ability of amiRNA vectors using *Agrobacterium*-mediated transient expression system.

2   Materials (*See* Note 1)

2.1   Plant Materials

1. *Nicotiana benthamiana*. Grow *N. benthamiana* plants from seeds in soil pots under greenhouse growth conditions.

2.2   amiRNA Design

1. Computer with internet access.
2. Vector NTI software (Life Technologies) (*see* **Note 2**).

2.3   Construction of amiRNA Shuttle Vector

1. *E. coli* DH5α competent cell.
2. 2× KaPa HiFi HotStart Ready Mix (Kapa Biosystems).
3. Inverse PCR primer pairs (*see* **Note 3**).
4. PCR Thermocycler.
5. Plasmid pOT2-poly-Cis (*see* **Note 4**).
6. QIAquick Gel Extraction Kit (Qiagen).
7. QIAquick PCR Purification Kit (Qiagen).
8. Restriction endonuclease SwaI.
9. T4 DNA ligase.
10. TAE (50×): 2 M Tris-acetate, 50 mM EDTA (pH 8.0).

2.4   Cloning of amiRNA Expression Cassette in Plant Binary Vector

1. *Agrobacterium tumefaciens* GV3101/C58C1 (pMP90) competent cells.
2. *E. coli* DH5α competent cell.
3. Plasmid pPZP212.
4. QIAquick Gel Extraction Kit.
5. QIAquick PCR Purification Kit.

6. Restriction endonuclease XbaI.

7. Restriction endonuclease XmaI.

8. 70 % (v/v) glycerol, Sterilize by autoclaving.

9. T4 DNA ligase.

2.5 Agrobacterium-Mediated Transient Expression

1. 0.4 M acetosyringone stock: Prepare the stock with DMSO, make aliquots and store at –20 or –70 °C until use.

2. 1 M 2-(N-morpholino)ethanesulfonic acid (MES) stock, pH 5.6.

3. 0.5 M MgCl$_2$ stock.

4. 20 µg/mL rifampin stock. Prepare the stock with methanol.

5. 100 µg/mL spectinomycin stock. Prepare the stock with water and sterilize with 0.2 µm sterilizing-grade filter.

6. YEB medium: 1.0 g/L of yeast extract, 5.0 g/L of beef extract, 5.0 g/L of peptone, 5.0 g/L of sucrose, 0.493 g/L of MgSO$_4$. Adjust pH to 7.2. Sterilize by autoclaving.

7. Infiltration medium: 10 mM MES, pH 5.6; 10 mM MgCl$_2$; 200 µM acetosyringone. Prepare fresh infiltration medium for agro-infiltration.

8. *N. benthamiana* plants, 4–5 weeks old.

9. Syringe, 3 mL.

10. Syringe needle.

2.6 Small RNA Extraction

1. Chloroform.

2. 75 % (v/v) ethanol.

3. 20 mg/mL glycogen, RNase free (Optional).

4. Isopropyl alcohol.

5. Liquid nitrogen.

6. TRIzol reagent (Life Technologies, USA).

7. 50 mL Falcon conical centrifuge tube.

8. Mortar and pestle.

9. Polypropylene centrifuge tube.

2.7 Small RNA Northern Blot

1. 40 % acrylamide–bis-acrylamide (19:1).

2. 25 % (w/v) ammonium persulfate (AP).

3. 0.5 M EDTA, pH 8.0.

4. 10 mg/mL salmon sperm DNA, sheared (Life Technologies).

5. 10 % (w/v) SDS.

6. Tetramethylethylenediamine (TEMED).

7. Urea.

8. 50× Denhardt's solution: 1 % bovine serum albumin (Fraction V; Sigma), 1 % Ficoll (Type 400, Pharmacia), 1 % polyvinyl-pyrrolidone (PVP).

9. Maleic acid buffer: 0.1 M maleic acid, 0.15 M NaCl, pH 7.5.

10. Washing buffer: 0.3 % (v/v) Tween 20 in maleic acid buffer.

11. Blocking solution: Dissolve blocking reagent 1 % (w/v) in maleic acid buffer under constantly stirring on a heating block at 65 °C. The solution remains opaque.

12. Antibody solution: Centrifuge the antibody for 5 min at 10,000 ×g in the original vial prior to each use, and pipette necessary amount carefully from the surface. Dilute anti-DIG-AP 1:10,000 (75 mU/mL) in blocking solution.

13. Detection buffer: 0.1 M Tris–HCl, 0.1 M NaCl, pH 9.5. CDP-Star working solution: Dilute CPD-Star 1:100 in Detection Buffer.

14. DNA 5′ or 3′-end labeling reagents. For 5′-end labeling, [γ-^{32}P] ATP, T4 polynucleotide kinase (PNK). For 3′-end labeling, digoxigenin-11-2′, 3′-dideoxy-uridine-5′-triphosphate (DIG-11-ddUTP), Terminal Deoxynucleotidyl Transferase (TdT).

15. Hybridization solution: 50 % formamide, 5× SSPE, 5× Dehardt's reagent, 0.5 % SDS, 80 μg/mL of salmon sperm DNA.

16. Small RNA loading buffer (2×): 98 % (w/v) deionized formamide, 10 mM EDTA pH 8.0, 0.025 % (w/v) bromophenol blue.

17. 20× SSC: 3.0 M NaCl, 0.3 M sodium citrate. Adjust pH to 7.0. Sterilize by autoclaving.

18. 20× SSPE: 3.0 M NaCl, 0.3 M NaH$_2$PO$_4$·H$_2$O, 0.02 M EDTA. Adjust pH to 7.4. Sterilize by autoclaving.

19. 10× TBE: 0.89 M Tris base, 0.89 M boric acid, 20 mM EDTA (pH 8.0).

20. CL-1000 UV Crosslinker (UVP Inc.).

21. Classic Blue X-Ray Film (Midwest Scientific Inc.).

22. Hybond-N$^+$ nylon membrane (GE Healthcare).

23. Phosphorimager screen.

24. Polyacrylamide gel electrophoresis (PAGE) apparatus: Mini-Protein 3 Cell System (Bio-Rad) and PowerPac Power Supply (Bio-Rad).

25. Trans-Blot SD Semi-Dry Transfer Cell (Bio-Rad).

26. T4 PNK (10,000 U/mL, Life Technologies).

27. γ-[^{32}P] ATP (10 μCi/μL, 3000 Ci/mmol, PerkinElmer).

3   Methods

3.1   amiRNA Design

1. Obtain the transcript sequence of the gene of interest (the target gene) and its isoforms from GenBank or other information sources (*see* **Note 5**).

2. Align the gene sequences for similarity analysis using Vector NTI (*see* **Note 2**).

3. According to the alignment analysis, identify a sequence of 21 nt in length as a candidate target of an amiRNA. The candidate target sequence can be specific to an individual gene of interest if it is to silence the individual gene or a common region of a group of gene isoforms if it is to simultaneously silence the gene group (*see* **Note 6**).

4. Input the selected target sequence (21-nt) into the miR168-based miRNA designer (http://www.smallrna.mtu.edu/ Tang_Website/submit.htm). The program will generate an amiRNA candidate that is perfectly base-paired (complementary) to the 21-nt target sequence, and then design a passenger strand (amiRNA*) according to the candidate amiRNA. The amiRNA/amiRNA* pair will have mismatched bases at positions 1 and 12 of the candidate amiRNA from 5′ to 3′ direction. Follow the website instructions to analyze the thermodynamic stability of the candidate amiRNA/amiRNA* duplex. An acceptable amiRNA candidate should meet the following criteria: the amiRNA is perfectly or almost perfectly base-paired with the targeting 21-nt sequence of the target gene or gene group; the nucleotide at position 1 from the 5′-end of amiRNA is an uracil (U) residue; and the combined free energy of the position 1–5 of the 5′-end of the amiRNA are higher than that of the amiRNA* (*see* **Note 7**).

5. Perform off-target assessment. When an amiRNA candidate is determined in **step 4**, conduct a blast search of the whole genome sequence of the species of the target gene or the EST database of the species. Avoid an amiRNA candidate that is perfectly paired at the seed sequence region (positions 2–8 of the amiRNA) with a potential off-target gene sequence in the genome or ESTs of the plant species.

6. After an acceptable amiRNA is determined, a pair of PCR primers (forward and reverse) for inverse PCR amplification of plasmid pOT2-poly-Cis will be designed by the web designer. An example of a pair of designed PCR primers, amiRPPO1234-FP/amiRPPO1234-RP (forward primer/reverse primer) is as follows (5′–3′direction): gcc<u>atttaaat</u>ctttattggtttgtgagcagggattg-g**TGTGAATGCAAATGAGTTTGT**atcggatcctcgaggtgtaaa/ gcc<u>atttaaat</u>gtcacgaggctgtgtcagccgaattgg**TGTGAATGC-GAATGAGCTTGA**atccgagcccgatggtgagac. The   upper-case

letters (5′ TGTGAATGCAAATGAGTTTGT 3′) in the forward primer represent an artificial sequence of amiRNA passenger strand (amiRNA*), and the bold upper-case letters (5′ **TGTGAATGCGAATGAGCTTGA 3′**) in the reverse primer represent a designed amiRNA target sequence, which is perfectly complementary to the designed amiRNA (the guide strand, which sequence is 5′ UCAAGCUCAUUCGCAUUCACA 3′). A restriction site of endonuclease SwaI (atttaaat, underlined) is introduced in both forward and reverse primers for easy cloning. The example amiRNA was designed to simultaneously knockdown four PPO genes, *StuPPO1*, *StuPPO2*, *StuPPO3*, and *StuPPO4* [5].

3.2 Construction of amiRNA Shuttle Vector by Inverse PCR

Designed amiRNA and amiRNA* are introduced in pri-miR168a to replace the sequence of miR168a and miR168a* by one-step inverse PCR amplification of plasmid pOT2-poly-Cis.

1. Prepare a PCR reaction for a total volume of 25 μL with 12.5 μL of 2× KaPa HiFi HotStart ReadyMix, 1 μL of 10 μM forward primer, 1 μL of 10 μM reverse primer, 1 μL of 2 ng/μL pOT2-poly-Cis plasmid, and 9.5 μL of nuclease-free water.

2. Run PCR on a PCR Thermocycler with an initial denaturation at 95 °C for 3 min, followed by 30 cycles of denaturation at 97 °C for 30 s, primer annealing at 58–63 °C for 30 s, and primer extension at 72 °C for 2 min, with a final 10 min extension at 70 °C.

3. Perform an 0.8 % agarose gel electrophoresis buffered in 1× TAE to separate the PCR product, excise the gel slice containing the DNA fragment of 3538 bp under UV light, and extract the DNA using a gel extraction kit, such as QIAquick Gel Extraction Kit.

4. Digest the gel-purified DNA fragment with endonuclease SwaI for 3 h at 25 °C, clean the digestion reaction using PCR purification kit, such as QIAquick PCR Purification Kit, and then do a self-ligation using T4 DNA ligase.

5. Transform the ligation reaction into appropriate *E. coli* competent cells, such as DH5α.

6. Pick transformed bacterial colonies, grow overnight at 37 °C, and extract plasmids.

7. Analyze the extracted plasmids by SwaI digestion. Positive plasmids will be digested by SwaI. Sequence the plasmid that can be digested by SwaI to confirm the intended construction and the sequence of the designs.

3.3 Cloning of amiRNA Expression Cassette in Plant Binary Vector

1. Double-digest the constructed amiRNA shuttle vector with restriction endonucleases XbaI and XmaI. Separate the digestion products by agarose gel electrophoresis and excise agarose gel slice containing the DNA fragment of 1933 bp,

namely P35S-amiR/*MIR168a*-T35S. The cassette of P35S-amiR/*MIR168a*-T35S contains CaMV 35S promoter, miR168a gene primary transcript inserted with the designed amiRNA duplex sequence, and CaMV 35S terminator. Extract the DNA fragment from the gel slice using a gel extraction kit, such as QIAquick Gel Extraction Kit.

2. Double-digest plant binary vector pPZP212 with XbaI and XmaI. Purify the double-digested vector using a PCR extraction kit, such as QIAquick PCR Purification Kit.

3. Ligate the cassette of the P35S-amiR/*MIR168a*-T35S into the XbaI/XmaI-digested pPZP212 vector using T4 DNA ligase.

4. Transform the ligation mixture into appropriate *E. coli* competent cells, such as DH5α.

5. Pick transformed bacterial colonies, grow overnight at 37 °C, and extract plasmids.

6. Analyze the isolated plasmids by DNA restriction analysis and identify the positive pPZP212-derived plasmid with the insert of the P35S-amiR/*MIR168a*-T35S cassette.

7. Transform the positive plasmid into an *Agrobacterium tumefaciens* strain, such as GV3101/C58C1 (pMP90) at 26–28 °C.

8. Pick transformed Agrobacterium colonies and grow overnight at 26–28 °C (*see* **Note 8**).

9. Make aliquots of agrobacterial glycerol stock by mixing 0.4 mL of the overnight culture and 0.4 mL of 70 % sterile glycerol in a 1.5 mL microcentrifuge tube. Store the stocks at –80 °C until use.

3.4 Agrobacterium-Mediated Transient Expression of amiRNA in N. benthamiana

1. Inoculate 3 mL of YEB medium containing 100 μg/mL spectinomycin and 20 μg/mL rifampin with a scoop of –80 °C agrobacterial stock containing the amiRNA binary vector using a bacterial inoculation loop, and grow overnight in an incubation shaker at 200 rpm and 26–28 °C.

2. Inoculate 0.2 mL of the overnight culture to a 20 mL of YEB medium containing 100 μg/mL spectinomycin, 20 μg/mL rifampin, 10 mM MES, and 20 μM acetosyringone in a 125-mL flask, and grow the medium overnight in an incubation shaker at 200 rpm and 26–28 °C.

3. Harvest cells by centrifuging the overnight culture at $4000 \times g$, 4 °C for 10 min.

4. Resuspend agrobacteria in freshly prepared infiltration medium, and adjust the OD_{600} of the agrobacteria suspension to 1.0. Incubate the cells at room temperature for at least 3 h prior to infiltration.

5. Make a tiny notch on a leaf surface of a 4–5-week-old *N. benthamiana* plant by gently touching the leaf surface with a needle. Be careful not to pierce through the leaf bottom. Support the leaf with one hand from under, and gently but firmly inject agrobacteria cell suspension into the intercellular space of the leaf through the notch using a 3 mL syringe without a needle. Infiltrate two fully expanded leaves per plant, and three to five plants for each treatment.

6. Collect 0.2 g of the infiltrated leaf tissue at 1–5 days post infiltration (dpi). Store the materials at –80 °C until use (*see* **Note 9**).

3.5 Small RNA Extraction Using TRIzol Reagent (See Note 10)

1. Weigh 0.2 g of leaf tissue (remove midrib), and saturate the material in liquid nitrogen. Grind the material to fine powder using mortar and pestle. Quickly transfer the powder to a cooled Falcon 50 mL conical centrifuge tube.

2. Add 2 mL of TRIzol reagent. Vortex vigorously to mix well, and incubate the sample for 5–10 min at room temperature to permit dissociation of nucleoprotein complexes.

3. Transfer the sample to a polypropylene centrifuge tube. Centrifuge at $12,000 \times g$, 4 °C for 15 min.

4. Transfer equal volumes of the cleared homogenate solution to two 1.5 mL microcentrifuge tubes (about 1.0 mL/tube).

5. Add 0.2 mL of chloroform to each tube (0.2 mL of chloroform per 1.0 mL of TRIzol reagent); cap the sample tube securely. Shake the tubes vigorously by hand for 15 s, and incubate the mixture at 15–30 °C for 2–3 min.

6. Centrifuge the tubes at no more than $12,000 \times g$, 4 °C for 15 min. Transfer the upper colorless aqueous phase to two new 1.5 mL microcentrifuge tubes (RNA remains exclusively in the aqueous phase. The volume of the aqueous phase is about 60 % of the volume of the Trizol Reagent used for homogenization, which is about 0.6 mL aqueous phase).

7. Add 1 μL of 20 mg/mL glycogen (RNase free) to each tube. Optional, if your RNA is of limited amount.

8. Add 0.6 mL of isopropyl alcohol to each tube. Mix well and incubate overnight at –20 °C to precipitate RNA.

9. Vortex briefly, centrifuge at $20,000 \times g$, 4 °C for 20 min.

10. Keep RNA pellet. Wash it once with 1 mL of 75 % ethanol, vortex briefly and spin at maximum speed, 4 °C for 10 min.

11. Keep RNA pellet. Dry RNA pellet briefly with speed vacuum.

12. Dissolve the RNA pellet with 100 μL of nuclease-free water. Measure the RNA concentration, and store the RNA at –80 °C until use.

**3.6   Analysis
of amiRNA Expression
by RNA Northern Blot**

1. Prepare a 17.5 % denatured polyacrylamide gel with 1.5 mm thickness as follows: In a Falcon 50 mL conical centrifuge tube, add 4.2 g of urea, 4.4 mL of 40 % acrylamide–bis-acrylamide (19:1) and 0.5 mL of 10× TBE. Mix well by vortex and microwave for 20 s to dissolve urea. Vortex again and heat again if required (urea will denature over 65 °C. Do not overheat.). Add sterile deionized water to bring volume up to 10 mL. Cool the solution on ice to slow down polymerization before adding 35 μL of 25 % (w/v) freshly made ammonium persulfate (AP) and 3.5 μL of TEMED. Mix well and immediately pour gel in a Mini-Protein 3 Cell System in a chemical hood. The gel will polymerize in about 30 min.

2. Assemble gel with electrophoresis unit. Load the gel tank with 0.5× TBE, clean the wells with pipette, and pre-run the gel for 30 min at 80 V.

3. Wash wells once again with 0.5× TBE before loading samples. It is critical to have clean wells to get proper sample loading.

4. Sample preparation. To the nuclease-free water dissolved RNA, add an equal volume of 2× small RNA loading buffer, mix well, spin down quickly at maximum speed in a benchtop centrifuge, heat in block at 75–95 °C for 5 min, and spin down again.

5. Load approximately 10 μg of total RNA per well. Run gel for approximately 4 h at 80 V or until bromophenol blue dye reaches the bottom of the gel.

6. Transfer the RNA from the gel to Hybond-N⁺ nylon membrane using 0.5× TBE and semidry blotting, such as Bio-Rad Trans-Blot Semi-Dry Transfer Cell.

7. Rinse the membrane with 100 mL of 2× SSPE, pH 7.4.

8. Crosslink the membrane at 120 mJ/cm² for 1 min in a CL-1000 UV-Crosslinker, and then bake the membrane at 80 °C for 30 min.

9. Place membrane in a hybridization bottle with the RNA side facing the interior, add 10 mL of hybridization solution and pre-hybridize in a hybridization oven with rotation control at 37 °C for at least 2 h.

10. Prepare probe. The probe is a set of synthesized 21-mer DNA oligonucleotide but labeled with radioisotope, dioxigenin or biotin at either the 5′ or the 3′ end. The nucleotide sequence of the probe should be reverse-complementary to the designed amiRNA sequence. We usually label a DNA oligo at the 5′-end using [γ-³²P] ATP and T4 polynucleotide kinase (PNK) or at the 3′-end using digoxigenin-11-2′, 3′-dideoxy-uridine-5′-triphosphate (DIG-11-ddUTP) and Terminal Deoxynucleotidyl Transferase (TdT). Two protocols are suggested here. Method

1, 5′-end labeling with [γ-^{32}P] ATP and T4 PNK: in a 1.5 mL microcentrifuge tube, mix 15.5 μL of H_2O, 5 μL of 5× T4 PNK forward reaction buffer, 2.5 μL of γ-[^{32}P] ATP (10 μCi/μL, 3000 Ci/mmol), 1 μL of 10 μM DNA oligo, and 1 μL of T4 PNK (10,000 U/mL), and incubate at 37 °C for 1 h. Method 2, 3′-end labeling with DIG-11-ddUTP and TdT: in a 1.5 mL microcentrifuge tube, mix 4.75 μL of H_2O, 2 μL of 5× TdT reaction buffer, 0.25 μL of DIG-11-ddUTP, 2.5 μL of 10 μM DNA oligo, and 0.5 μL of TdT (15 U/μL), and incubate at 37 °C for 30–60 min.

11. After pre-hybridization for a minimum of 2 h, carefully add the probe, either 25 μL of the γ-[^{32}P] labeled, or 1 μL of the DIG labeled (1:10,000), from **step 10** to the hybridization solution without directly touching the membrane and continue the hybridization overnight at 37 °C (*see* **Note 11**).

12. For detection using radioisotope, follow **steps 13–16**.

13. Remove the hybridization solution and keep in a Falcon 50 mL conical centrifuge tube for possible reuse or discard it into radioactive waste.

14. Rinse membrane immediately with 10 mL of 2× SSC and 0.1 % SDS at 37 °C for 5 min.

15. Wash membrane in 25 mL of 2× SSC at 37 °C twice, each for 25 min. Check the background signal after each wash.

16. Wrap membrane with plastic wrap and expose overnight or longer to a Phosphorimager screen.

17. For detection using DIG detection system, follow **steps 18–27** (*see* **Note 12**).

18. Remove the hybridization solution. Rinse the membrane immediately with 20 mL of 2× SSC and 0.1 % SDS at 37 °C for 5 min.

19. Wash membrane in 25 mL of 2× SSC at 37 °C twice, each for 25 min.

20. Rinse membrane in 25 mL of Washing buffer for 1–5 min.

21. Incubate membrane for 45 min in 25 mL of Blocking solution.

22. Incubate membrane for 45–60 min in 15 mL of Antibody solution.

23. Wash membrane in 25 mL of Washing buffer four times, 10 min per wash.

24. Transfer membrane to a small tray and equilibrate it in 25 mL of Detection buffer for 2–5 min.

25. Place membrane with RNA side facing up on a transparent sheet of plastic and cover the membrane surface with 1 mL of

CDP-Star working solution. Incubate for 3 min. Flip the RNA membrane upside-down on the solution and incubate for another 2 min.

26. Squeeze out excess liquid from the membrane, but not let it completely dry. Wrap membrane with plastic wrap.

27. Expose to X-ray film for 3 h or overnight at 15–25 °C before development of the film (*see* **Note 13**).

---

4   Notes

1. Some routine molecular biology laboratory equipment, solutions, and kits, such as pipettes and pipette tips, centrifuge, microcentrifuge tubes, agarose gel electrophoresis apparatus and running solutions, plasmid extraction kits, UV image system, are not listed.

2. Researchers may wish to use their favorite sequence analysis software, such as DNASTAR Lasergene, CLC Genomics.

3. Design of the inverse PCR primers is introduced in Subheading 3.1. We have DNA oligo primers synthesized by Integrated DNA Technologies (USA). Although the primers have more than 80-mer, PAGE or HPLCpurification of the oligos is generally not necessary.

4. Plasmid pOT2-poly-Cis was originally constructed in Dr. Guilaing Tang's Laboratory [1, 20].

5. Isoforms or analogs of the gene of interest can be retrieved by blast search of the NCBI or other genome sequence or EST database.

6. We normally look for a 21-nt sequence with an adenine (A) residue at the proximal 3′ end (position 21 in a 5′–3′ direction), and no other specific nucleotide base requirement is asked for selecting of the target sequence.

7. For designing one amiRNA to target multiple genes, G:U wobbling base-pairing can be considered if perfect complementarity between candidate amiRNA and the target sequence is not possible. In addition, mismatches between amiRNA and its target are permitted at position 1, 20, and 21 of the amiRNA (counted from 5′ to 3′ direction) but it may reduce the efficiency of RNAsilencing [10, 21].

8. Transformation of a pPZP-derived plasmid to *Agrobacterium* is normally successful with appropriate antibiotics selection, but we usually conduct a PCR detection of the transformed *Agrobacterium* overnight culture to confirm the existence of the pPZP-derived plasmid before making agrobacterial glycerol stock.

9. Transient expression of the specific amiRNAs would decrease but small interference RNA (siRNA) level caused by RNA silencing would increase over time [1]. For better representation of the specific amiRNA expression level, collect and detect leaf samples between 24 and 48 h post infiltration.

10. Commercial small RNA isolation kits, such as NucleoSpin miRNA Kit (Macherey-Nagel) are available for small RNA isolation, but not examined in our assay.

11. We do not purify the labeled probes, but purification of the probes using G25 spin column may be considered.

12. The DIG detection procedures (**steps 20–23**) can be done in either a hybridization bottle or a small tray container at 37 °C or room temperature.

13. Besides specific expression of amiRNAs, RNAsilencing is also induced by *Agrobacterium*-mediated transient expression system over time [22, 23]. In addition to the detection of the specifically expressed amiRNAs, this Northern blot may also detect siRNAs that have nucleotide sequence homology with the amiRNAs. However, hybridization signals for the siRNAs are usually weak at 1–2 days post-infiltration. Another noticeable pattern is that siRNAs may be detected with two band sizes (~22- and 24-nt) [1]. For this reason, we suggest that an amiRNA vector with a weak hybridization signal may not be a good vector for stable plant genetic transformation.

References

1. Bhagwat B, Chi M, Su L, Tang H, Tang G, Xiang Y (2013) An in vivo transient expression system can be applied for rapid and effective selection of artificial microRNA constructs for plant stable genetic transformation. J Genet Genomics 40:261–270

2. Niu QW, Lin SS, Reyes JL, Chen KC, Wu HW, Yeh SD, Chua NH (2006) Expression of artificial microRNAs in transgenic *Arabidopsis thaliana* confers virus resistance. Nat Biotechnol 24:1420–1428

3. Ossowski S, Schwab R, Weigel D (2008) Gene silencing in plants using artificial microRNAs and other small RNAs. Plant J 53:674–690

4. Schwab R, Ossowski S, Riester M, Warthmann N, Weigel D (2006) Highly specific gene silencing by artificial microRNAs in Arabidopsis. Plant Cell 18:1121–1133

5. Chi M, Bhagwat B, Lane WD, Tang G, Su Y, Sun R, Oomah BD, Wiersma PA, Xiang Y (2014) Reduced polyphenol oxidase gene expression and enzymatic browning in potato (*Solanum tuberosum* L.) with artificial microRNAs. BMC Plant Biol 14:62

6. Deveson I, Li J, Millar AA (2013) MicroRNAs with analogous target complementarities per-

form with highly variable efficacies in Arabidopsis. FEBS Lett 587:3703–3708

7. Li JF, Chung HS, Niu Y, Bush J, McCormack M, Sheen J (2013) Comprehensive protein-based artificial microRNA screens for effective gene silencing in plants. Plant Cell 25:1507–1522

8. Eamens AL, McHale M, Waterhouse PM (2014) The use of artificial microRNA technology to control gene expression in *Arabidopsis thaliana*. Methods Mol Biol 1062:211–224

9. Li JF, Zhang D, Sheen J (2014) Epitope-tagged protein-based artificial miRNA screens for optimized gene silencing in plants. Nat Protoc 9:939–949

10. Tang G, Reinhart BJ, Bartel DP, Zamore PD (2003) A biochemical framework for RNA silencing in plants. Genes Dev 17:49–63

11. Schwartz T, Blobel G (2003) Structural basis for the function of the beta subunit of the eukaryotic signal recognition particle receptor. Cell 112:793–803

12. Khvorova A, Reynolds A, Jayasena SD (2003) Functional siRNAs and miRNAs exhibit strand bias. Cell 115:209–216

13. Schwarz DS, Hutvagner G, Du T, Xu Z, Aronin N, Zamore PD (2003) Asymmetry in the assembly of the RNAi enzyme complex. Cell 115:199–208

14. Guo Y, Han Y, Ma J, Wang H, Sang X, Li M (2014) Undesired small RNAs originate from an artificial microRNA precursor in transgenic petunia (*Petunia hybrida*). PLoS One 9(6):e98783

15. Park W, Zhai J, Lee JY (2009) Highly efficient gene silencing using perfect complementary artificial miRNA targeting AP1 or heteromeric artificial miRNA targeting AP1 and CAL genes. Plant Cell Rep 28:469–480

16. Ai T, Zhang L, Gao Z, Zhu CX, Guo X (2011) Highly efficient virus resistance mediated by artificial microRNAs that target the suppressor of PVX and PVY in plants. Plant Biol 13:304–316

17. Alvarez JP, Pekker I, Goldshmidt A, Blum E, Amsellem Z, Eshed Y (2006) Endogenous and synthetic microRNAs stimulate simultaneous, efficient, and localized regulation of multiple targets in diverse species. Plant Cell 18:1134–1151

18. Kung YJ, Lin SS, Huang YL, Chen TC, Harish SS, Chua NH, Yeh SD (2012) Multiple artificial microRNAs targeting conserved motifs of the replicase gene confer robust transgenic resistance to negative-sense single-stranded RNA plant virus. Mol Plant Pathol 13:303–317

19. Qu J, Ye J, Fang R (2007) Artificial microRNA-mediated virus resistance in plants. J Virol 81:6690–6699

20. Tang G, Yan J, Gu Y, Qiao M, Fan R, Mao Y, Tang X (2012) Construction of short tandem target mimic (STTM) to block the functions of plant and animal microRNAs. Methods 58:118–125

21. Haley B, Zamore PD (2004) Kinetic analysis of the RNAi enzyme complex. Nat Struct Mol Biol 11:599–606

22. Johansen LK, Carrington JC (2001) Silencing on the spot. Induction and suppression of RNA silencing in the Agrobacterium-mediated transient expression system. Plant Physiol 126:930–938

23. Voinnet O, Rivas S, Mestre P, Baulcombe D (2003) An enhanced transient expression system in plants based on suppression of gene silencing by the p19 protein of tomato bushy stunt virus. Plant J 33:949–956

Chapter 15

Knockdown of Polyphenol Oxidase Gene Expression in Potato (*Solanum tuberosum* L.) with Artificial MicroRNAs

Ming Chi, Basdeo Bhagwat, Guiliang Tang, and Yu Xiang

Abstract

It is of great importance and interest to develop crop varieties with low polyphenol oxidase (PPO) activity for the food industry because PPO-mediated oxidative browning is a main cause of post-harvest deterioration and quality loss of fresh produce and processed foods. We recently demonstrated that potato tubers with reduced browning phenotypes can be produced by inhibition of the expression of several PPO gene isoforms using artificial microRNA (amiRNA) technology. The approach introduces a single type of 21-nucleotide RNA population to guide silencing of the PPO gene transcripts in potato tissues. Some advantages of the technology are: small RNA molecules are genetically transformed, off-target gene silencing can be avoided or minimized at the stage of amiRNA designs, and accuracy and efficiency of the processes can be detected at every step using molecular biological techniques. Here we describe the methods for transformation and regeneration of potatoes with amiRNA vectors, detection of the expression of amiRNAs, identification of the cleaved product of the target gene transcripts, and assay of the expression level of PPO gene isoforms in potatoes.

Key words Artificial microRNA (amiRNA), Gene knockdown, Polyphenol oxidase (PPO), Potato genetic transformation, RNA silencing, *Solanum tuberosum*, StuPPO

1 Introduction

Plant polyphenol oxidases (PPOs) are a class of copper-binding enzymes responsible for oxidative browning reactions, a main cause of quality loss in fresh produce and processed foods [1, 2]. PPOs are encoded by a gene family composed of multiple members of highly conserved genes in many plant species [2, 3]. A survey of the potato genome revealed nine PPO-like gene models, namely *StuPPO1* to *StuPPO9* [4]. Analysis of the expression sequence tag (EST) data of the gene models indicates *StuPPO1* to *StuPPO4* are the four major PPO genes that are actively expressed in different potato tissues, whereas *StuPPO5* to *StuPPO8* are under-transcribed in potatoes, and *StuPPO9* is more likely an inducible PPO gene expressed in response to disease defense. Making crop varieties

Arthur Germano Fett-Neto (ed.), *Biotechnology of Plant Secondary Metabolism: Methods and Protocols*, Methods in Molecular Biology, vol. 1405, DOI 10.1007/978-1-4939-3393-8_15, © Springer Science+Business Media New York 2016

with low browning reactions is of great interest to the food industry. One strategy is to suppress PPO activities by knocking down PPO gene expression in the plant tissues.

Reports have described methods of suppression of PPO gene expression using genetic engineering of PPO gene fragments in configurations of sense, antisense or double-stranded RNA in crops [5–8]. The approaches functioned through establishment of an RNA silencing mechanism guided by a population of heterogeneous small interfering RNAs (siRNAs). In addition, we recently demonstrated that PPO gene isoforms in potato can be knocked down individually or in combination using artificial microRNA (amiRNA) technology [4]. A series of potato tubers with reduced browning phenotypes resulted from the amiRNA-mediated suppression of the expression of different potato PPO gene isoforms. The greatest reduction of PPO activity and browning reaction in the potato tubers required simultaneous inhibition of the gene expressions of *StuPPO1* to *StuPPO4*. The method utilizes microRNA (miRNA) gene backbone to produce amiRNAs, an artificially selected small RNA of 21 nucleotides (nt) in length.

The amiRNAs then guide RNA silencing complex to silence the gene of interest in the engineered plant species [9–11]. This strategy provides an option to transform plants with very small RNA molecules, and generates only a single type of 21-nt RNA population all with the same selective nucleotide sequence in the transgenic plants. In addition, possible off-targets of an amiRNA candidate can be easily assessed by blast-search of the related plant genome sequences at the amiRNA designing stage. Thus the potential risk of off-target RNA silencing is greatly minimized. It is also more feasible for specific silencing of an individual gene or simultaneously silencing several or all genes of a multigene family. Importantly, accuracy and efficiency of the processes can be detected and monitored at every step using molecular biological methods, as we describe in this chapter.

2 Materials

2.1 Plant Genetic Transformation and Plant Tissue Culture

All plant tissue culture media (BM, MS0, CC, SM1, SM2, and SM3) are adjusted to pH 5.8 and autoclaved at 103 pKa and 121 °C for 20 min. Growth regulators are added after autoclaving. Stock solutions of growth regulators are prepared in DMSO (*see* **Note 1**). Stock solutions of antibiotics are filter-sterilized and added to cooled, but not gelled media following autoclaving.

1. Potato tissue culture plants, 3 to 4-week old.
2. Basic Medium (BM): 1× Murashige and Skoog (MS) basal medium, vitamin mixture, 20 g/L of sucrose and 7 g/L of agar.
3. MS0 Medium (MS0): 1× MS basal medium, no agar, no hormones or antibiotics.

4. Co-Cultivation medium (CC): BM plus 0.2 mg/L NAA, 0.02 mg/L GA3, 2.5 mg/L zeatin riboside.

5. Stage 1 Selection Medium (SM1): BM plus 0.2 mg/L NAA, 0.02 mg/L GA3, 2.5 mg/L zeatin riboside, 7 mg/L kanamycin, and 300 mg/L of cefotaxime.

6. Stage 2 Selection Medium (SM2): BM plus 0.02 mg/L NAA, 0.02 mg/L GA3, 2 mg/L zeatin riboside, 50 mg/L kanamycin, and 300 mg/L of cefotaxime.

7. Stage 3 Selection Medium (SM3): BM plus 100 mg/L kanamycin and 300 mg/L cefotaxime.

8. Luria Broth (LB) media: 10 g/L tryptone, 10 g/L yeast extract, 10 g/L NaCl. Adjust pH to 7.5. For solid media, 18 g/L agar is used.

9. Rifampin, 20 μg/mL. Prepare the stock with methanol.

10. Spectinomycin, 100 μg/mL. Prepare the stock with water and sterilize with 0.2 μm sterilizing-grade filter.

11. *Agrobacterium tumefaciens* strain GV3101/C58C1 (pMP90) transformed with pPZP212 derived amiRNA vector (Table 1) (*see* **Note 2**).

2.2 Small RNA Extraction

1. Chloroform.

2. 75 % (v/v) ethanol.

3. 20 mg/mL glycogen, RNase free (Optional).

4. Isopropyl alcohol.

5. Liquid nitrogen.

6. TRIzol reagent (Life Technologies, USA).

7. 50 mL Falcon conical centrifuge tube.

8. Mortar and pestle.

9. Polypropylene centrifuge tube.

2.3 Detection of amiRNA Expression

1. Agarose gel electrophoresis apparatus.

2. TAE stock solution (50×): 2 M Tris-acetate, 50 mM EDTA (pH 8.0).

3. PCR primers for detection of the designed amiRNAs (oligo1, 3, 4, 6, 7 and ath5.8S, Table 2) (*see* **Note 3**).

4. NCode miRNA first-strand cDNA synthesis kit (Life Technologies).

5. 2× Taq MasterMix (Applied Biological Materials, Canada).

6. PCR thermocycler.

7. TRIzol reagent (Life Technologies).

Table 1
List of the amiRNA vectors, amiRNA names, and amiRNA target genes

Construct[a]	amiRNA name	amiRNA sequence (5′→3′)	Target gene (GenBank ID)	Comment
pPZamiRPPO1	amiRPPO1	UUGGUGACUGGUGCAAUUGAC	*StuPPO1/POTP1/P2* (XM_006355177/M95196/M95197)	Knockdown *StuPPO1* effectively
pPZamiRPPO3	amiRPPO3	UUGUUCACUGGGGGGAGUGUA	*StuPPO3/POT33* (XM_006365320/U22922)	Knockdown *StuPPO3* effectively
pPZamiRPPO23	amiRPPO23	UCAUCAACUGGAGUUGAGUUG	*StuPPO2/POT32* (U22921); *StuPPO3/POT33*	Knockdown *StuPPO2* and *StuPPO3* effectively, moderately knockdown *StuPPO4*
pPZamiRPPO234A	amiRPPO234A	AAGAACUCGGAGUUCAACCAA	*StuPPO2/POT32*; *StuPPO3/POT33*; *StuPPO4/POT72* (U22923)	Knockdown *StuPPO2*, *StuPPO3* and *StuPPO4* effectively, moderately knockdown *StuPPO1*
pPZamiRPPO1234	amiRPPO1234	UCAAGCUCAUUCGCAUUCACA	*StuPPO1/POTP1/P2*; *StuPPO2/POT32*; *StuPPO3/POT33*; *StuPPO4/POT72*	Knockdown *StuPPO1, 2, 3 and 4* effectively

[a]Only the vectors that were proven to be effective in knockdown of PPO gene expressions are listed in Table 1 [4]

Table 2
List of the primers used for detection of potato amiRNAs and the cleaved potato PPO gene products

Name	Sequence (5′–3′)	Comment
oligo1	TTGGTGACTGGTGCAATTGAC	Forward primer for detection of amiRPPO1
oligo3	TTGTTCACTGGGGGGAGTGTA	Forward primer for detection of amiRPPO3
oligo4	TCATCAACTGGAGTTGAGTTG	Forward primer for detection of amiRPPO23
oligo6	AAGAACTCGGAGTTCAACCAA	Forward primer for detection of amiRPPO234A
oligo7	TCAAGCTCATTGCATTCACA	Forward primer for detection of amiRPPO1234
ath5.8S	ACGTCTGCCTGGGTGTCACAA	Forward primer for detection of 5.8S rRNA (internal control)
oligo8	ACCARAGTYACCGCAATAGT	Common primers for synthesis of the cDNA from 3′ end of PPO gene transcripts
oligo9	TCKTCATCTTCCAAWCCAWTRTCC	Common reverse primer for 1st-round PCR
oligo10	TTRTCAAGCTCATYCGCATTCACA	Common reverse primer for nested-PCR-1
oligo11	ACGAAGCTGGTCTGGTGATAGAGA	Specific reverse primer for detection of *StuPPO1* gene mRNA in nested-PCR-2
oligo12	CATATCTTATGCTACTGAATGTCAAC	Specific reverse primer for detection of *StuPPO2* gene mRNA in nested-PCR-2
oligo13	TGAAGCAGGCCTATTGATGGAAAAT	Specific reverse primer for detection of *StuPPO3* gene mRNA in nested-PCR-2
oligo14	TCATACTTTATTTCATTGAACGTTAGC	Specific reverse primer for detection of *StuPPO4* gene mRNA in nested-PCR-2

Note: K stands for nucleotide G or T; R for A or G; W for A or T; Y for C or T

2.4 Detection of Cleaved mRNA Products

1. Mortars and pestles.
2. Liquid nitrogen.
3. Dynabeads mRNA purification kit (Life Technologies).
4. FirstChoice RLM-RACE kit (Life Technologies).
5. MinElute PCR purification kit (Qiagen).
6. QIAquick gel extraction kit (Qiagen).
7. QIAquick PCR purification kit (Qiagen).
8. pGEM-T Easy Vector system (Promega).
9. Spectrum plant total RNA kit (Sigma-Aldrich).
10. TRIzol reagent (Life Technologies).
11. 2× Taq MasterMix (Applied Biological Materials, Canada).
12. PCR thermocycler.
13. Primers (DNA oligos) for detection of PPO gene products (oligo 8-14, Table 2).
14. TBE stock solution (5×): 0.45 M Tris-borate, 10 mM EDTA (pH 8.0).

2.5 Analysis of PPO Gene Expression

1. Mortars and pestles.
2. Liquid nitrogen.
3. Spectrum plant total RNA kit (Sigma-Aldrich).
4. TURBO DNase and buffer (Life Technologies).
5. RNeasy mini kit (Qiagen).
6. RT primer mix (provided in QuantiTech reverse transcription kit, Cat. No. 205310, Qiagen) (*see* **Note 4**).
7. RNAseOUT Recombinant ribonuclease inhibitor (40 U/μL) (Life Technologies).
8. SuperScript III reverse transcriptase (200 U/μL) and reagents (Life Technologies).
9. EvaGreen 2× qPCR MasterMix-iCycler (Applied Biological Materials, Canada).
10. qPCR primers for reference genes (Table 3) (*see* **Note 5**).
11. qPCR primers for PPO genes (Table 3) (*see* **Note 6**).
12. 96-well microplates for qPCR.
13. CFX96 Real-Time PCR detection system.

3 Methods

3.1 Genetic Transformation and Regeneration of Potato Transgenic Plants

The potato transformation protocol utilizes internodal explants obtained from aseptically grown (in vitro) plantlets and modified from the protocols as described [12]. Maintain stock cultures of potato plantlets in vitro in Magenta GA7 vessels or test tubes (*see* **Note 7**). Prior to genetic transformation, subculture the potato

Table 3
qRT-PCR primers used in quantification of PPO gene expression in potato tuber tissue

Primer pair name	Gene name	Gene ID	Primer sequences (forward/reverse, 5'–3')	Amplicon length (bp)	Amplification efficiency (%) ± S.D.	R^2
cyclo1/2	Cyclophilin	AF126551	CTCTTCGCCGATACCACTCC/TCACACGGTGGAAGGTTGAG	121	0.984±0.006	0.9999±0.0001
ef2/3	ef1α	AB061263	ATTGGAAACGGATATGCTCCA/TCCTTACCTGAACGCCTGTCA	101	0.873±0.004	0.9996±0.0004
ppo35/36	*StuPPO1/POTP1/2*	M95196/M95197	GACCAGCTTCGTCAAGGACTA/TTGTCAACGTTCAGGAACACA	121	0.998±0.003	0.9927±0.0003
ppo41/42	*StuPPO2/POT32*	U22921	ATATCGCGACTGTTGATTTCC/GTCGCACCTTCAATGGAGATA	133	1.012±0.006	0.9966±0.0003
ppo46/47	*StuPPO3/POT33*	U22922	ATGGCGTAACTTCAAACCAAA/CCATCTTCGTGAGTGGGAATA	98	0.992±0.006	0.9954±0.0004
ppo54/55	*StuPPO4/POT72*	U22923	TCTGGTGCCAAAGAAAGGTAA/ACAAACAATCCGCAGATTCAA	96	1.000±0.001	0.9997±0.0001
ppo31/32	*StuPPO1/POTP1/2*	M95196/M95197	TTGACACACCTCAGCTTCAGA/GTAAGCAGCACCGAAGAATTG	104	0.954±0.001	0.9999±0.0001
ppo40/38	*StuPPO2/POT32*	U22921	CGATGTAACACGTGACCAAAG/GTTTCAACTTCATTGCCGAAA	75	0.9313±0.003	0.9998±0.0001
ppo48/49	*StuPPO3/POT33*	U22922	GGACAAGGTCCAAACAACTCA/CTTCGACGTTATTCCCAAGAA	134	0.924±0.008	0.9994±0.0004
ppo52/53	*StuPPO4/POT72*	U22923	ATTGAATCTGCGGATTGTTTG/CAAGGAAATTGGATTCATTCG	108	0.926±0.004	0.9994±0.0001

plantlets using single-node explants and culture for 4 weeks. Excise the upper six internodes and use them as explants for inoculation with *Agrobacterium tumefaciens*. Shoots are propagated on basic medium (BM) without any growth regulators. Cultures are incubated under photoperiod of 16 h light (100 µE/m²/s)/8 h dark at 22 °C.

1. Initiate a 5 mL liquid LB culture of *Agrobacterium tumefaciens* strain GV3101/C58C1 (pMP90) containing amiRNA binary vector from a glycerol stock or plate culture, with appropriate antibiotics (*see* **Note 8**). Grow the culture overnight at 28 °C on an incubation shaker at 200 rpm.

2. Transfer the 5 mL culture to 50 mL of LB, plus antibiotics, in a 250 mL flask. Grow the culture at 28 °C on an incubation shaker at 200 rpm until an OD_{550} of 1.2 is achieved.

3. Transfer the culture to a sterile centrifuge tube and spin at $3000 \times g$ for 10 min at 4 °C to harvest the cells. Resuspend the bacterial cells in 50 mL liquid MS0 medium.

4. Prepare internode explants from 3- to 4-week-old stock plants. Explants should be approximately 10 mm length and over 2.5 mm wide. Inoculate 50 explants with 25 mL of *Agrobacterium* suspension in a 9 cm diameter petri dish. Incubate for 30 min at room temperature.

5. Following incubation, pour off *Agrobacterium* suspension and blot explants with sterile filter paper. Plate explants onto co-cultivation medium (CC). Seal plates and co-cultivate for 3 days at 22 °C and low irradiance (20 µE/m²/s).

6. After the co-cultivation, wash the explants with liquid MS medium (MS0) containing 1 g/L of cefotaxime for 30 min.

7. Transfer explants to Stage 1 Selection Medium (SM1) (*see* **Note 9**). Place ten explants per petri dish. Seal plates and incubate for 2 weeks with photoperiod of 16 h light (100 µE/m²/s)/8 h dark at 22 °C.

8. Transfer explants to SM2 and incubate for 2 weeks with photoperiod of 16 h light (100 µE/m²/s)/8 h dark at 22 °C.

9. After 2 weeks on SM2, transfer explants to SM3. Thereafter, continue to transfer to fresh SM3 every 2 weeks until shoot regeneration ceases.

10. Shoot regeneration will start approximately 3 weeks after the time explants are placed on selection medium. Remove green shoots larger than 1 cm and individually culture in test tubes with SM3 (*see* **Notes 10** and **11**).

11. Screen well-rooted plantlets to confirm that they are transgenic. Root formation from the cut ends of shoots under a selection pressure of 100 mg/L of kanamycin is an indication

that the shoots/plantlets are transgenic. However, before plantlets are taken to the greenhouse, their status should be confirmed by PCR.

12. Subculture transgenic plants to provide sufficient replicates for transferring to greenhouse and for making an in vitro repository (*see* **Note 12**). One transgenic event should be cultured per tube.

13. Remove plantlets at the 4–5 node stage, wash off agar from the roots and plant in moist potting mix in 5 L pots (*see* **Note 13**). Grow in a greenhouse with photoperiod of 16 h light (150 µE/ m^2/s)/8 h dark at 20 °C. Maintain humidity at 100 % and reduce it to ambient levels gradually over a 1–2 week period. Fertilize at weekly intervals with a liquid fertilizer.

14. Harvest tubers at the end of the growing period (*see* **Note 14**).

3.2 Small RNA Extraction Using TRIzol Reagent (See Note 15)

1. Weigh 0.2 g of leaf tissue (remove midrib), and saturate the material in liquid nitrogen. Grind the material to fine powder using mortar and pestle. Quickly transfer the powder to a cooled Falcon 50 mL conical centrifuge tube.

2. Add 2 mL of TRIzol reagent. Vortex vigorously to mix well, and incubate the sample for 5–10 min at room temperature to permit dissociation of nucleoprotein complexes.

3. Transfer the sample to a polypropylene centrifuge tube. Centrifuge at 12,000×g, 4 °C for 15 min.

4. Transfer equal volumes of the cleared homogenate solution to two 1.5 mL microcentrifuge tubes (about 1.0 mL/tube).

5. Add 0.2 mL of chloroform to each tube (0.2 mL of chloroform per 1.0 mL of TRIzol reagent); cap the sample tube securely. Shake the tubes vigorously by hand for 15 s, and incubate the mixture at 15–30 °C for 2–3 min.

6. Centrifuge the tubes at no more than 12,000×g, 4 °C for 15 min. Transfer the upper colorless aqueous phase to two new 1.5 mL microcentrifuge tubes (RNA remains exclusively in the aqueous phase. The volume of the aqueous phase is about 60 % of the volume of the TRIzol Reagent used for homogenization, which is about 0.6 mL aqueous phase).

7. Add 1 µL of 20 mg/mL glycogen (RNase free) to each tube. Optional, if RNA is of limited amount.

8. Add 0.6 mL of isopropyl alcohol to each tube. Mix well and incubate overnight at –20 °C to precipitate RNA.

9. Vortex briefly, centrifuge at 20,000×g, 4 °C for 20 min.

10. Keep RNA pellet. Wash it once with 1 mL of 75 % ethanol, vortex briefly and spin at maximum speed, 4 °C for 10 min.

11. Keep RNA pellet. Dry RNA pellet briefly with speed vacuum.

12. Dissolve the RNA pellet with 100 μL of nuclease-free water. Measure the RNA concentration, and store the RNA at –80 °C until use.

3.3 Detection of amiRNA Expression by Reverse Transcription PCR (RT-PCR)

Poly(A) tailing of total RNA and synthesis of cDNAs are performed using the NCode miRNA first-strand cDNA synthesis kit following the manufacturer's instructions but with some modifications.

1. To add poly(A) tails to total RNA, dilute 1 μL of 10 mM ATP to 4 μL of 1 mM Tris-HCl, pH 8.0 (Diluted ATP).

2. Assemble the reaction mixture: 1 μg of total RNA, 2 μL of 5× miRNA reaction buffer, 1 μL of 25 mM $MnCl_2$, 1 μL of the diluted ATP, 0.2 μL of Poly(A) Polymerase, and DEPC-treated water to a final volume of 10 μL.

3. Incubate at 37 °C for 15 min.

4. To synthesize first-strand cDNA, add 4 μL of the polyadenylated RNA from **step 2** in a new microcentrifuge tube with 1 μL of annealing buffer and 3 μL of 25 μM Universal RT Primer.

5. Incubate at 65 °C for 5 min.

6. Place the tube on ice for 1 min.

7. Add 10 μL of 2× first-strand reaction mix and 2 μL of SuperScript III/RNaseOut enzyme mix to the tube.

8. Incubate at 50 °C for 50 min, and then incubate at 85 °C for 5 min to stop the reaction.

9. To amplify amiRNA by PCR, prepare PCR reaction mixture: Mix 0.5–1 μL of the cDNA from **step 3**, 5 μL of 1.6 mM of the amiRNA specific primer (forward primer, Table 2, *see* **Note 3**), 5 μL of 1.6 mM of the universal qPCR primer (reverse primer, *see* **Note 3**), 10 μL of 2× PCR Taq MasterMix (ABM) in a 0.2 mL PCR tube.

10. Run PCR on a PCR thermocycler, such as MyCycler with an initial denaturation step at 94 °C for 1 min, followed by 30–40 cycles of denaturation at 94 °C for 30 s, primer annealing at 57–63 °C for 15 s, and primer extension at 72 °C for 20 s, with a final 10-min extension at 70 °C.

11. Separate PCR products by electrophoresis in 3 % agarose gel buffered in 1× TAE. Stain the gel with ethidium bromide. A DNA fragment of ~65 bp (including amiRNA and adaptor) will be amplified in positive RNA samples by RT-PCR (*see* **Note 16**).

3.4 Detection of Cleaved mRNA Products of amiRNA Target (5′ RACE-PCR)

The cleaved mRNA product of the amiRNA target in the transgenic potatoes can be detected using the FirstChoice RACE Kit with some modifications. The kit includes a procedure to remove the free 5′-phosphates from molecules such as the 3′ end fragment of a cleaved mRNA initially using Calf-intestinal alkaline phosphatase (CIAP). The cap structure of the capped mRNA (full-length mRNA) will then be removed using Tobacco acid pyrophosphatase (TAP). The decapped mRNA is then ligated with a designed 5′ RACE RNA adapter using T4 RNA ligase. In our application, we directly ligate the cleaved fragment to the 5′ RACE RNA adapter. The CIAP and TAP treatments are omitted to prevent capped-mRNA from ligating to the adapter. Thus, only cleaved RNA fragment can be amplified by the following PCR reactions [4]. Cleaved mRNA is usually less stable due to the subsequent degradation processes in the cell. The low to rare amount of the cleaved mRNAs increases the difficulty to clone the specifically cleaved fragment by a simple RACE-PCR. Hence, multiple rounds of nested-PCR are designed here for improving the sensitivity of the detection.

1. Disrupt potato tissue in liquid nitrogen using mortar and pestle. Extract total RNA using TRIzol reagent or Spectrum plant total RNA kit (*see* **Note 17**).

2. Extract poly(A)$^+$ RNA from total RNA using the Dynabeads mRNA purification kit according to the manufacturer's instructions.

3. Add 5′ RACE adaptor to RNA: Mix 2 μg of poly(A)$^+$ RNA, 1 μL of 5′ RACE Adapter, 1 μL of 10× RNA Ligase Buffer, 2 μL of T4 RNA Ligase (2.5 U/μL), and nuclease-free water to a final volume of 10 μL. Incubate at 37 °C for 1 h.

4. Synthesize first-strand cDNA: mix 2 μL of ligated RNA from **step 3**, 4 μL of dNTP Mix (2.5 mM each dNTP), 1 μL of primer oligo8 (2 μM) (Table 2), 2 μL of 10× RT Buffer, 1 μL of RNase Inhibitor (10 U/μL), 1 μL of M-MLV reverse transcriptase, and nuclease-free water to a final volume of 20 μL. Incubate at 42 °C for 1 h.

5. Purify the first-strand cDNA reaction using MinElute PCR purification kit and elute the cDNA in 80 μL of nuclease-free water (*see* **Note 18**).

6. For the first-round PCR, prepare PCR reaction mixture: Mix 8 μL of the cDNA from **step 5**, 1 μL of primer oligo9 (40 μM) (Table 2), 1 μL of 5′ RACE Outer Primer (10 μM, provided in the FirstChoice RACE Kit), and 10 μL of 2× PCR Taq MasterMix in a 0.2 mL PCR tube (*see* **Note 19**).

7. Run PCR on a PCR thermocycler, such as MyCycler with an initial denaturation step at 94 °C for 1 min, followed by 35 cycles of denaturation at 94 °C for 30 s, primer annealing at 55 °C for 30 s, and primer extension at 72 °C for 45 s, with a final 10-min extension at 70 °C.

8. Check the amplification products by running 5 μL of the PCR reaction in a 1.8 % agarose gel buffered in 0.5× TBE.

9. Purify the remaining first-round PCR product using QIAquick PCR Purification Kit and elute the DNA in 120 μL of nuclease-free water (*see* **Note 18**).

10. For the nested-PCR-1, prepare PCR reaction mixture: Mix 0.5 μL of the purified PCR product from **step 7**, 1 μL of primer oligo10 (20 μM) (Table 2), 1 μL of 5′ RACE Inner Primer (10 μM, provided in the FirstChoice RACE Kit), 10 μL of 2× PCR Taq MasterMix and 7.5 μL of nuclease-free water in a 0.2 mL PCR tube (*see* **Note 19**).

11. Run PCR on a PCR thermocycler, such as MyCycler with an initial denaturation step at 94 °C for 1 min, followed by 35 cycles of denaturation at 94 °C for 30 s, primer annealing at 55 °C for 30 s, and primer extension at 72 °C for 35 s, with a final 10-min extension at 70 °C.

12. Check the amplification products by running 5 μL of the PCR reaction in a 1.8 % agarose gel buffered in 0.5× TBE.

13. Purify the remaining nested-PCR-1 product using QIAquick PCR Purification Kit and elute the DNA in 120 μL of nuclease-free water (*see* **Note 18**).

14. For the nested-PCR-2, prepare four different PCR reaction mixtures: Mix 0.5 μL of the purified PCR product from **step 9**, 1 μL of specific primer (oligo11, oligo12, oligo13, or oligo14, Table 2) (10 μM), 1 μL of 5′ RACE Inner Primer (10 μM, provided in the FirstChoice RACE kit), 10 μL of 2× PCR Taq MasterMix (ABM), and 7.5 μL of nuclease-free water in a 0.2 mL PCR tube (*see* **Note 19**).

15. Run PCR on a PCR thermocycler, such as MyCycler with an initial denaturation step at 94 °C for 1 min, followed by 35 cycles of denaturation at 94 °C for 30 s, primer annealing at 60 °C for 30 s, and primer extension at 72 °C for 30 s, with a final 10 min extension at 70 °C.

16. Run PCR product in 1.8 % agarose gel buffered in 0.5× TBE. Stain the gel with ethidium bromide to examine the specific PCR amplification.

17. Isolate the specific amplified DNA band from the agarose gel. Purify the DNA fragment using QIAquick gel extraction kit. Clone the DNA in pGEM-T easy vector system. Isolate and sequence the plasmids.

3.5 Analysis of PPO Gene Expression by Real-Time Quantitative Reverse Transcription PCR (qRT-PCR)

1. Disrupt potato tissues in liquid nitrogen using mortar and pestle. Extract total RNA from the tissues using Spectrum plant total RNA kit according to the manufacturer's instruction and elute the total RNA using 100 μL elution solution (*see* **Note 20**).

2. Remove the trace amounts of DNA from the purified total RNA samples: Mix 88 μL of the purified total RNA from the

step 1, 10 µL of 10× TURBO DNase Buffer and 2 µL of TURBO DNase (2 U/µL). Incubate the mixture at 37 °C for 45 min, and then re-purify the reactions using RNeasy Mini Kit. Determine the concentration and purity of the re-purified total RNA samples using ND-1000 spectrophotometer.

3. To synthesize first-strand cDNA, mix 1 µg of re-purified total RNA, 1 µL of RT primer mix (provided in the QuantiTect reverse transcription kit, Ca. No. 205310) (*see* **Note 4**), 1 µL of dNTP Mix (10 mM) and nuclease-free water to 13 µL. Incubate the mixture at 65 °C for 5 min and then cool down on ice for 1 min.

4. To the mixture above, add 4 µL of 5× First-Strand buffer, 1 µL of 0.1 M DTT, 1 µL of RNase Inhibitor, 1 µL of SuperScript III reverse transcriptase (200 U/µL), and nuclease-free water to a final volume of 20 µL. Mix well. Incubate the mixture at 25 °C for 10 min, followed by 42 °C for 55 min and then at 70 °C for 15 min. Store the cDNAs at –20 °C before use (*see* **Note 21**).

5. Set up qRT-PCR reactions in a 96-well microplate. Each reaction volume is 20 µL and prepared as follows: 10 µL of Eva green master mix (Applied Biological Materials Inc.), 0.5 µM of forward and reverse primers (primer pair mix, 5 µL) and 5 µL of appropriately diluted cDNA template or nuclease-free water for no-template control (NTC). Each reaction is set up in triplicate.

6. Perform the qRT-PCR reactions in a CFX96 real-time PCR detection system with the following program: one initial denaturation cycle at 95 °C for 10 min, followed by 40 cycles at 95 °C for 30 s and at 60 °C for 30 s. Fluorescence signal is measured at the end of each annealing and extension step at 60 °C. At the end of the qRT-PCR run, perform a melting curve analysis with a temperature gradient of 0.1 °C /s from 60 to 95 °C (*see* **Note 22**).

7. Data analysis: Calculate the relative expression levels of the PPO gene expressions based on a method described in Pfaffl (2001). Perform data normalization using the gene expression values of the ef1α and cyclophilin in the samples.

4   Notes

1. Stock solutions of most plant growth regulators can be prepared with DMSO. By so doing, there is no need for filter-sterilization. However, the minimum volume of DMSO should be used, that is convenient to dispense and to minimize any possible negative effects of DMSO on the growth of plant tissue.

2. The amiRNAs were designed and constructed in *Arabidopsis thaliana* miR168a gene backbone and plant binary vector pPZP212 according to the principles and methods described in [13]. The efficiencies of knockdown of PPO gene expressions of the amiRNA vectors listed in Table 1 were examined and presented in [4].

3. The sequences of the forward PCR primers (oligo1, 3, 4, 6, and 7 in Table 2) are identical to the sequences of the designed 21-nt amiRNAs listed in Table 1 except that the PCR primers are synthesized as DNA oligos. Oligo ath5.8S is designed to detect 5.8S ribosomal RNA as an internal control. The reverse primer is the universal qPCR primer provided in the NCode miRNA First-strand cDNA synthesis kit.

4. The RT primer mix provided in QuantiTech reverse transcription kit (Qiagen) is a mixture of oligo-dT and random primers. cDNAs synthesized with this RT primer mix provided more repeatable qRT-PCR results in quantification of potato PPO gene expressions in our experiments [4] (M. Chi et al., unpublished).

5. The qPCR primer pairs (cyclo1/2 and ef2/3) are the same as the designs by Nicot et al. [14]. PCR amplification specificity of the primer pairs was examined by gel electrophoresis, amplicon sequencing and post-PCR melting curve analysis.

6. PCR amplification specificity of the qPCR primer pairs listed in Table 3 was examined by gel electrophoresis, amplicon sequencing and post-PCR melting curve analysis. In total, 30 pairs of PCR primers were synthesized and screened. Only eight primer pairs, two primer pairs for each PPO gene, which passed the specificity examinations, are listed in Table 3. Although two primer pairs for each PPO gene, i.e., *StuPPO1*, *StuPPO2*, *StuPPO3*, or *StuPPO4* are provided in Table 3, quantification assay of each gene expression can be done with only one of the two primer pairs.

7. In vitro cultures can be established from sprouting tubers or greenhouse-grown plants [15]. Culture vessels used should allow for ventilation. This would prevent the accumulation of ethylene in the vessels.

8. Antibiotics used in this step are rifampin (20 μg/L) and spectinomycin (100 μg/L).

9. Stage 1 Selection Medium (SM1) contains 7 mg/L kanamycin. Explants are transferred to this medium immediately after co-cultivation. We have found that a lower level of selection at an earlier stage reduces the incidence of escapes while at the same time does not reduce regeneration as compared to a higher selection concentration (B. Bhagwat et al., unpublished).

10. Putative transgenic shoots should be removed together with a piece of the internodal explant from which it regenerated. This approach prevents the regeneration of the "same transgenic event."

11. By culturing putative transgenic shoots individually in tubes, removal for greenhouse growth is easier.

12. (Optional) An early-stage screening of the PPO gene knockdown transgenic plants from the cultured replicates using qRT-PCR is suggested because it would reduce workload for making the transgenic plants. The qRT-PCR method is the same as described in Subheading 3.5 in this chapter.

13. Alternatively, in vitro plantlets may be first established in vivo using multi-cell seedling trays. This allows for a more convenient handling of the hardening-off process.

14. Tubers harvested can be stored under controlled environment conditions for long durations. Alternatively, in vitro cultures can be allowed to produce micro tubers which can then be maintained for long durations under controlled environment conditions.

15. Commercial small RNA isolation kits, such as NucleoSpin miRNA kit (Macherey-Nagel) are available for small RNA isolation.

16. Because the size of the Poly(A) tail is not indicated in the NCode miRNA first-strand cDNA synthesis kit (Life Technologies) by the manufacturer, the size of a positive amplification was estimated from our positive control.

17. Potato leaf or potato tuber cortex (a narrow range of 2 mm below the skin) tissues are suggested for extraction of total RNA because of higher PPO gene transcripts in these tissue types (M. Chi et al., unpublished).

18. This cleanup step is optional, but it can reduce nonspecific amplifications in the follow-up PCR.

19. Setup of a control PCR sample that contains only the common forward primer but no reverse primer is a good option. The control will be helpful to identify the nonspecific PCR amplifications.

20. For early-stage screening of the PPO gene knockdown transgenic plants from the tissue cultures, extraction of total RNA from the whole plant (leaf and stem tissues) of the cultured replicates is suggested. For detection of PPO gene expression in potato tubers, extraction of total RNA from potato tuber cortex tissue is suggested (*see* **Note 17**).

21. Dilute each cDNA reaction with 180 μL of nuclease-free water, make aliquots and store at −20 °C until use.

22. Perform the melting curve analysis to ensure that only single specific product is produced.

References

1. Holderbaum DF, Kon T, Kudo T, Guerra MP (2010) Enzymatic browning, polyphenol oxidase activity, and polyphenols in four apple cultivars: dynamics during fruit development. HortScience 45:1150–1154

2. Mayer AM (2006) Polyphenol oxidases in plants and fungi: going places? A review. Phytochemistry 67:2318–2331

3. Tran LT, Taylor JS, Constabel CP (2012) The polyphenol oxidase gene family in land plants: lineage-specific duplication and expansion. BMC Genomics 13:395

4. Chi M, Bhagwat B, Lane WD, Tang G, Su Y, Sun R, Oomah BD, Wiersma PA, Xiang Y (2014) Reduced polyphenol oxidase gene expression and enzymatic browning in potato (*Solanum tuberosum* L.) with artificial microRNAs. BMC Plant Biol 14:62

5. Bachem CW, Speckmann G-J, van der Linde PC, Verheggen FT, Hunt MD, Steffens JC, Zabeau M (1994) Antisense expression of polyphenol oxidase genes inhibits enzymatic browning in potato tubers. Nat Biotechnol 12:1101–1105

6. Coetzer C, Corsini D, Love S, Pavek J, Tumer N (2001) Control of enzymatic browning in potato (*Solanum tuberosum* L.) by sense and antisense RNA from tomato polyphenol oxidase. J Agric Food Chem 49:652–657

7. Murata M, Nishimura M, Murai N, Haruta M, Homma S, Itoh Y (2001) A transgenic apple callus showing reduced polyphenol oxidase activity and lower browning potential. Biosci Biotechnol Biochem 65:383–388

8. Richter C, Dirks ME, Gronover CS, Prüfer D, Moerschbacher BM (2012) Silencing and heterologous expression of ppo-2 indicate a specific function of a single polyphenol oxidase isoform in resistance of dandelion (*Taraxacum officinale*) against *Pseudomonas syringae* pv. tomato. Mol Plant Microbe Interact 25:200–210

9. Bhagwat B, Chi M, Su L, Tang H, Tang G, Xiang Y (2013) An in vivo transient expression system can be applied for rapid and effective selection of artificial microRNA constructs for plant stable genetic transformation. J Genet Genomics 40:261–270

10. Ossowski S, Schwab R, Weigel D (2008) Gene silencing in plants using artificial microRNAs and other small RNAs. Plant J 53:674–690

11. Schwab R, Ossowski S, Riester M, Warthmann N, Weigel D (2006) Highly specific gene silencing by artificial microRNAs in Arabidopsis. Plant Cell 18:1121–1133

12. Beaujean A, Sangwan RS, Lecardonnel A, Sangwan-Norreel BS (1998) *Agrobacterium-mediated* transformation of three economically important potato cultivars using sliced internodal explants: an efficient protocol of transformation. J Exp Bot 49:1589–1595

13. Bhagwat B, Chi M, Han D, Tang H, Tang G, Xiang Y (2015) Design, construction and validation of artificial microRNA vectors using Agrobacterium-mediated transient expression system. Methods Mol Biol

14. Nicot N, Hausman JF, Hoffmann L, Evers D (2005) Housekeeping gene selection for real-time RT-PCR normalization in potato during biotic and abiotic stress. J Exp Bot 56:2907–2914

15. Millam S (2006) Potato (*Solanum tuberosum* L.). Methods Mol Biol 344:25–36

INDEX

Arthur Germano Fett-Neto (ed.), *Biotechnology of Plant Secondary Metabolism: Methods and Protocols*, Methods in Molecular
Biology, vol. 1405, DOI 10.1007/978-1-4939-3393-8, © Springer Science+Business Media New York 2016

Printed in the United States
By Bookmasters